The Pursuit of Quantum Gravity

Larry Murphy

Bryce De Witt

Cécile DeWitt-Morette

The Pursuit of Quantum Gravity

Memoirs of Bryce DeWitt from 1946 to 2004

Cécile DeWitt-Morette
Department of Physics
Center for Relativity
University of Texas at Austin
Austin Texas
USA
cdewitt@physics.utexas.edu

ISBN 978-3-642-14269-7 e-ISBN 978-3-642-14270-3
DOI 10.1007/978-3-642-14270-3
Springer Heidelberg Dordrecht London New York

Library of Congress Control Number: 2011921724

Cover design: WMXDesign GmbH, Heidelberg

Printed on acid-free paper

Springer is part of Springer Science+Business Media (www.springer.com)

*Dedicated to our daughters, Nicolette, Jan,
Christiane, Abigail*

Preface

This book is written for the curious reader. I hope it will also be a good read for the professional physicist. It develops and supplements the biographical memoir (16 pages) prepared by Steven Weinberg for the National Academy of Sciences.[1] It is also a companion to Bryce DeWitt's last book [BD 103][2] *The Global Approach to Quantum Field Theory* (1042 pages, first published in 2003, reprinted with corrections in 2004). Bryce's book is an excellent physicist's record of his pursuit of quantum gravity.

As an introduction to the theme of the book, two pieces from Bryce DeWitt are reproduced *in extenso*:

- "Why Physics?" consists of a few hand-written pages addressed to his grandson, and
- A posthumously published article "Quantum Gravity: Yesterday and Today" addressing the questions: "Why Quantum Gravity?" and "Why Link Quantum Gravity and Bryce DeWitt's Memoirs?"

The structures of quantum physics and gravitation physics do not constitute a single logical system. Digging deeper into their respective structures may reveal a common root. Therefore the pursuit of quantum gravity requires a profound understanding of both quantum physics and gravitation physics. As G.A. Vilkovisky wrote to me, "Quantum Gravity is a combination of two words, and one should know both. Bryce understood this as nobody else, and this wisdom is completely unknown to many authors of the flux of papers that we see nowadays."[3]

Sections I and II deal with the status of quantum physics and gravitation physics in the late forties. A couple of major topics are then selected in each

[1] Steven Weinberg, "*Bryce Seligman DeWitt, 1923–2004: A Biographical Memoir*", National Academy of Sciences (2008). This memoir incorporates materials given to Weinberg by DeWitt before his death and is reproduced in Sect. V.1.

[2] References to DeWitt's own list of publications are indicated in the text in square brackets prepended BD.

[3] E-mail correspondence dated Dec. 28, 2007.

field because they were of special interest to Bryce. Quantum gravity itself is the focus of Sect. III. It captures Bryce's work from 1946 to 2004 and places it in its scientific, historical, and human context.

I have been blessed and burdened by an enormous amount of material for the preparation of this book; in addition to, I believe, a complete set of published material, I have most of Bryce's reports to institutions and referee reports to journals, letters of recommendation, correspondence, research proposals, unpublished documents, and my memories (*les souvenirs que la mémoire invente*![4]).

I have included extensive quotes from Bryce. They are printed in blue. Bryce's writings were carefully crafted, and he could not stand any editing (except by his fourth daughter Abigail, who edited his last writing "God's Rays"). Therefore I could not summarize his quotes, paraphrase them, or select brief quotes out of context. I have added introductions, motivations, and historical comments that place his technical progress in a larger context and make his research papers easier to read.[5] I have included a few technical notes when necessary to justify some statements. The reader not interested in technical details can gloss over them without losing the thread of the topic. The material in blue (Bryce's quotes) and the material in black (Cécile's contributions and others') are meaningful by themselves; together they make a whole. Including extensive quotes occasionally creates a few duplications, but they are not *verbatim* duplications; a later quote shows the maturation of an earlier one.

My task has been made manageable thanks to the help of Brandon DiNunno, John Stachel, and Lawrence Shepley, and to the offer from The University of Texas Center for American History to be the repository of Bryce's documents.[6]

The support that The University of Texas provides to its Professors Emeriti is gratefully acknowledged. Without an office, and all the facilities of the Physics Department, I could not have completed this task in a timely fashion. In addition, my office is close to Austin Gleeson's and whenever I needed something, I only had to poke my head in his always open door.

I have enjoyed many visits to the *Institut des Hautes Etudes Scientifiques* at Bures-sur-Yvette, France, an oasis in the fast moving, crowded schedule of modern times – an oasis where one finds the peace necessary for concentration and the stimulation of other travellers following their own pursuits.

---

[4] Recollections that memory invents.

[5] For instance, in Bryce's condensed notation, an index may refer to coordinates both in the domain and in the range of the function. Occasionally he used mathematical terms heuristically; I have often been frustrated by his use of the word "measure" in functional integrals. Like many physicists he talks of gauge fields as connections. In reality gauge potentials are pull-backs of connections on a principal bundle.

[6] *see* Sect. V.4.

These pages are memoirs around the theme "Quantum Gravity".[7] Once a neglected problem in the backwaters of the flow of physics research, quantum gravity has now become a major challenge in fundamental physics.[8]

The pursuit of quantum gravity goes on. This book began as a scrapbook of its beginnings and matured into its present form, though it is not, and was not meant to be, a scholarly piece of Science History in the strict sense. Some colleagues have suggested pertinent additions, in particular Thibault Damour, Jürg Fröhlich, Phillip Morrison, and Larry Smarr, but I have kept in mind the size of the book and its scientific level – and hope to have achieved a nicely readable balance.

There are some who seem to question the rôle of publishers in the age of the internet; but a great publisher eliminates the gangue from the mineral ore by addition or by subtraction. My manuscript has been made into a book thanks to the care and expertise of Dr. Christian Caron, Executive Publishing Editor at Springer-Verlag, Theodor C.H. Cole, Copy and Language Editor, and Birgit Münch, Desk Editor.

Cécile Morette, épouse DeWitt[9]

[7] For a balanced non-technical review on the development of the field as a whole, *see* e.g. C. Rovelli, Notes for a brief history of quantum gravity, arXiv: gr-qc/0006061v3 (23 Jan 2001).

[8] It is also mentioned outside physics. For instance, when Alan Sokal wanted to play a hoax on a cultural studies journal, he sent a parody of postmodern academic prose titled "Transgressing the Boundaries Towards a Transformative Hermeneutics of Quantum Gravity" to *Social Text*. The article was published in 1996, and the hoax became an often-quoted anecdote (Alan D. Sokal, "Transgressing the Boundaries Towards a Transformative Hermeneutics of Quantum Gravity", *Social Text* **46/47**, 217–252 (1996)).

[9] I use my French legal name to indicate that my work as a physicist and as a spouse are independent of each other. The concept of a person, or an object, assuming different functions in different contexts is not unusual. In mathematics an operator is defined only together with the space on which it operates. The same is true in human relationships. As a mother and as a wife a person's behavior may be different. The label "spouse" was inappropriately attached to my qualifications as physicist, and antinepotism regulations were invoked to deny me a position in the same department as my husband. The concept "conflict of interest" is sometimes used inappropriately. My French name, as recorded on my birth certificate is "Cécile, Andrée, Paulette Morette-Payen, veuve DeWitt" I prefer "épouse" to "veuve". French birth certificates are updated to reflect an individual's *État Civil.*

Contents

Introduction

Why Physics?

DeWitt wrote the following pages for his grandson, Benjamin Bryce Arar, probably in the mid-nineties.

" The word is Greek: Φυσικος, meaning the study of nature. Nature is everything that is. So that's a tall order. I don't remember exactly when I knew I wanted to be a physicist. At the age of five I certainly had never heard of the word "physics". Or if I had, it made zero impression. But I had a grandmother who was a Christian fundamentalist and who used to tell me that I would live to see the Messiah return to Earth in all his glory. That sounded neat – like science fiction. But I was curious to know how the descent from the sky could be managed. I didn't think it would be easy.

What do I remember of those early years? My grandparents were poor, and before they came to live with us they stayed in a ramshackle farmhouse on eighty acres of leased vineyard. The house had no electricity, and water only when the gasoline-powered pump was on. But I liked to stay with them. I liked it when the coal oil lamps were lit in the evenings and, after supper, when Grandfather would read out of the Bible: "The heavens declare the glory of God," or (the Lord speaking to Job) "Gird up now thy loins like a man: ... Where wast thou when I laid the foundations of the earth? ... When the morning stars sang together, and all the sons of God shouted for joy?" Grandfather was not a fundamentalist. He believed in evolution (to Grandmother's lifelong dismay). As a boy in Canada he had wanted to be an astronomer and had made his own telescope.

There were other things I liked about the farm. I liked to feed the chickens and fetch the eggs. And I liked to sit on one of the two holes in the outhouse, read the catalogue and think about the universe. At night I slept with a big old alarm clock. The hands and the numbers on the dial bore a paint containing radium, that made them luminous in the dark. When I

C. DeWitt-Morette, *The Pursuit of Quantum Gravity: Memoirs of Bryce DeWitt from 1946 to 2004*, DOI 10.1007/978-3-642-14270-3_1, © Springer-Verlag Berlin Heidelberg 2011

Fig. 1 *Why Physics* is addressed to his grandson, Benjamin (Ben) Bryce Arar. Bryce loved to read aloud. *Kim* and *The Elephant s Child* by Kipling were two of his favorites, and so was the *Cat in the Hat.*

held the clock up close to my face I could see the scintillations. I was puzzled by the flicker and didn't learn until years later that I was seeing the effects of individual particles emitted during radioactive decay.

In the San Joaquin valley where I grew up, the night sky was closer to the earth than it is in most places nowadays. The air was drier then, and there were far fewer bright lights. I remember running down the streets of our little town after dark, with my brother, and looking at the sky as we ran. It seemed as if we could fall right into the depths of space. And space *looked* deep. There was no impression of the stars being painted on the sky.

Since there was obviously so much more of space than there was of Earth, Space was where I wanted to be. When I was a little older I used to read science fiction avidly. It was a more optimistic and naive science fiction than what you find now – robots and space travel mostly. It was fun and allowed plenty of scope for imagination. But the best stories were those that took seriously the implacable constraints that Nature imposes: Conservation of energy. No Speeds faster than light. The vastness of interstellar distances. I have always been a thorough going realist as far as Nature is concerned, but delighted when the reality turns out to involve such fairy tales as quantum mechanics and curved spacetime.

When I was ten or eleven I thought I wanted to be an astronomer. It was not until I reached high school that I learned what physics was. I always liked math. In fact, I thought that the study of nature at the most fundamental level was math. I was wrong, but not very wrong. And that is something that is so deeply curious that many people have been intrigued by it. Math is *not* physics. But it is so close to physics that you can't separate the two. For example, complex numbers are a part of nature, and yet their properties do not depend on a single fact of nature. (I didn't learn about complex numbers until I was at least nine or ten – much older than you, Ben[10].)

Physicists have to be scrupulously honest, and mathematics helps them to be. But mathematics also provides beautiful frameworks for organizing their ideas. Physicists come in two types. There are those who get an intense joy out of tinkering with real things. They have learned how to build fantastic devices that, in the hands of engineers, have made it possible for us to compute rapidly, to communicate around the world and in deep space, and to look in minute detail inside the human body. These experimental physicists, as they are called, interact constantly with the other kind of physicists, the theorists, who are not so fond of tinkering and prefer instead to sit back and puzzle out what it all means. You might think that theorists are lazy, but they're not (at least the good ones aren't). The puzzles of Nature are often baffling. In biology the unravelling of these puzzles requires the patient cooperation and insights of hundreds of researchers, all doing hands-on work. But the deep insights of physics have usually come from just a few theorists, thinking hard for many years, using mathematical ideas in the most brazenly opportunistic ways, and never touching a piece of apparatus other than a pencil or a computer.

When I started graduate study (after a stint as a naval aviator to get space travel out of my system) I became acquainted with two fabulously beautiful ideas: General relativity and Quantum Field Theory. Neither idea can be expressed without abstract mathematics, but that's not why they are beautiful. They are beautiful because of the way they use the math and the way they fit with the physical ideas of earlier years. But I found that *they didn't fit with each other*. General relativity – Einstein's theory of gravity and curved space-time – was a great advance over Newton's theory, which had lasted for over two centuries, but there were only three tiny effects that it explained better. Quantum Field Theory, on the other hand, explained and predicted many

[10] A footnote from Ben: "When I was four years old, I once pressed "$\sqrt{-1}$" on a pocket calculator. The calculator screen told me, "Error", so I asked my mom what $\sqrt{-1}$ was. She told me to call Grandmama (Cécile). Grandmama said that $\sqrt{-1}$ was i, but that my calculator didn't know i because i was "imaginary". She then proceeded to explain some properties of complex numbers. I tried to ask complicated questions to stump her, but she seemed to know all the answers."

things in modern laboratories that could not be well understood or even anticipated before, none of which included gravity. Since Einstein's theory was a field theory, although a somewhat unusual one, I resolved to drag it forcibly into the mainstream of physics by quantizing it. The bulk of my professional life has been devoted to this effort in one way or another.

At the beginning my goal was regarded by colleagues as mildly indecent: not what "real" physicists do. It is ironic that today the effort is regarded as central to the whole of theoretical physics. Alas, the problem is still not solved, at least in any practically useful way, and there is another irony in this. My own efforts proved in the end to be more applicable to a different kind of field, known as a *gauge field*, of fundamental importance in its own right.

Theoretical physicists are the modern theologians. But they are amateurs. They are impressed, as anyone should be, by the scale and the astonishing properties of our universe, and they would like to see the face of God. When they are young they are set out, full of optimism, to discover how the universe ticks. Ultimately they learn, in the words of Steven Weinberg, that "the more the universe seems comprehensible, the more it seems pointless." At least it doesn't point to any goal for human beings. It is certainly not incompatible with human beings, for they are a part of it and can exploit it to their own advantage as far as they are able. But human beings must set their own goals. Among many worthy goals, one is to understand Nature. Another is to resist the temptation of believing that the universe exists for Man.

Weinberg explained later that what he meant by "pointless" was "that the universe itself suggests no point," in other words that the question Why does the universe exist? seems to have no answer. Yet there may be kind of an answer hidden in another puzzle I have often posed in the following question to pure mathematicians: Is mathematics the creation of human beings or is it simply there to be discovered, as a creation of God? Curiously, the greatest mathematicians almost invariably reply that it is there to be discovered. In other words, in any universe however bizarre, that is capable of producing intelligent beings, these beings would discover: the integers, real and complex numbers, symmetries, discrete and continuous groups, manifolds, and so on. This suggests that mathematical ideas exist independently of intelligent beings. They certainly exist independently of experimental facts. Do they really exist *a priori*? Can mathematics exist independently of a universe in which to express it? I think the answer is no. Physics and mathematics cannot be separated from one another any more than two quarks can be pulled apart, and this is reflected in the cosmos itself. There exist miniuniverses that are so simple and so trivial that they would indeed be pointless. But in our own universe intelligent beings may have a modest role to play after all."

Why Quantum Gravity? Why Link Quantum Gravity and Bryce DeWitt's Memoirs?

The following posthumously published article [BD 107][11] was found in DeWitt's files without an indication of its purpose. When Brandon DiNunno, then an undergraduate student, asked Cécile DeWitt for guidance in studying Quantum Gravity, she suggested that they look up references justifying comments made in this paper (There were no references in DeWitt's text). One computation was attributed to "a student of mine" (no name). Completed with 39 references and the name of the student, Walter Wesley[12], the article was sent to George Ellis, Editor-in-Chief with Hermann Nicolai, of *General Relativity and Gravitation*. He accepted it overnight and wrote a heartwarming abstract:

"Bryce DeWitt was one of the great pioneers of quantum gravity. This unpublished lecture gives his recent views on the topic, which we believe will be of great interest not only to researchers involved in modern attempts to quantize Einstein's theory, but also to a much wider audience. It is the first installment of a book *The Pursuit of Quantum Gravity 1946–2004; Memoirs of Bryce DeWitt* that Cécile DeWitt is preparing.We would like to thank her for the permission to publish this lecture separately in *General Relativity and Gravitation*. Readers who have unpublished material such as letters from Bryce, and would be willing to send copies to Cécile, are hereby invited to do so. She would be very grateful. G.F.R. Ellis, H. Nicolai (Editors-in-Chief)."

"A paper that I published in 1967 [1] began with these words: "Almost as soon as Quantum Field Theory was invented by Heisenberg, Pauli, Fock, Dirac, and Jordan, attempts were made to apply it to fields other than the electromagnetic field which had given it—and indeed quantum mechanics itself—birth. In 1930 Rosenfeld [2, 3] applied it to the gravitational field which, at the time, was still regarded as the *other* great entity of Nature. Rosenfeld was the first to note some of the special technical difficulties involved in quantizing gravity and made some early attempts to develop general methods for handling them. As an application of his methods he computed the gravitational self energy of a photon in the lowest order of perturbation theory. He obtained a quadratically divergent result, confirming that the divergence malady of field theory, which had already been discovered in connection with the electron's electromagnetic

[11] Bryce DeWitt "Quantum Gravity: Yesterday and Today." Edited by Cécile DeWitt-Morette and Brandon DiNunno. *General Relativity and Gravitation* **41–2**, 413–419 (2009). Errata: *General Relativity and Gravitation* **41–3**: 617 (2009). We gratefully acknowledge Springer-Verlag for allowing us to reproduce the article here in full.

[12] *see* Sect. III.4 the last entry in the list of Ph.D. graduates.

self-energy, was widespread and deep seated. It is tempting, and perhaps no longer premature, to read into Rosenfeld's result a forecast that quantum gravidynamics was destined, from the very beginning, to be inextricably linked with the difficult issues lying at the theoretical foundations of particle physics."

In 1967 such a forecast *was* premature, and yet any thoughtful person had to ask himself: What is the gravitational field doing there, in such splendid isolation? What if one simply dragged it forcibly into the mainstream of theoretical physics, and quantized it? In 1948 Julian Schwinger, my Ph.D. thesis advisor, gave me permission to reperform Rosenfeld's calculation, but in a manifestly gauge-invariant way, with the aim of showing that Rosenfeld's result implies merely a renormalization of charge rather than a nonvanishing photon mass. In fact, the one-loop result is nil, although to show this, using Schwinger's clumsy methods rather than Feynman's elegant diagrams, was not easy [4].

Most of you can have no idea how hostile the physics community was, in those days, to persons who studied general relativity. It was worse than the hostility emanating from some quarters today toward the string-theory community. In the mid fifties Sam Goudsmit, then Editor-in-Chief of the *Physical Review*, let it be known that an editorial would soon appear saying that the *Physical Review* and *Physical Review Letters*[13] would no longer accept "papers on gravitation or other fundamental theory." That this editorial did not appear was due to the behind-the-scenes efforts of John Wheeler.

It was possible, in those days, to divide theoretical physicists into two camps, according to their view of the role of gravitation in physics, just as it is possible to divide mathematicians into two camps according to their answer to the question: Is mathematics *there*, to be discovered, or is it a free invention of the human mind?[14] With very few exceptions mathematicians of world class stature say that it is there. Those with limited horizons – the second raters – say that it is a free invention.

Let me illustrate the situation in physics by the following anecdote: In November 1949, at the Institute for Advanced Study (where I was already beginning to cast glances at Cécile), I met Pauli. I was hoping to spend some time as a postdoc at the ETH, so Pauli asked me what I was working on. I said I was trying to quantize the gravitational field. For many seconds he sat silent, alternately shaking and nodding his head (a nervous habit he had, affectionately known as die Paulibewegung). He finally said "That is a very important problem—but it will take someone really smart!"

Neither Pauli nor Schwinger had limited horizons, nor had Feynman, who began to think about quantum gravity after attending a conference on the role of gravitation in physics organized by Cécile in January 1957 [5, 6].

[13] S. Goudsmit was also Managing Editor of *Reviews of Modern Physics*.

[14] *see* for instance: J.P. Changeux and A. Connes "Matière à Pensée" (O. Jacob 1992).

Feynman's thinking culminated in his discovery of ghosts, which he announced at a gravity conference in Warsaw in 1962 [7]. At that time he knew how to incorporate ghosts only into one-loop diagrams. The two-loop case, with its overlapping subgraphs, was unraveled by me in 1964 [8], by methods that would clearly give me the answer in any order. By the end of 1965 I was able to express the rules for ghosts to all orders in terms of a functional integral that could easily be shown to be invariant under deformations in gauge-breaking terms. These rules were written up and submitted to the Physical Review in 1966 [9–11].

Here I have to backtrack and describe the situation outside the physics community. In 1955 I received a letter from the Glenn L. Martin Aircraft Company which began with the words "It occurred to us a number of years ago that our company was vitally interested in gravity ... " They were looking for physicists who could build an antigravity device, and turned to me because I had won first prize in a Gravity Research Foundation essay contest. In those days, only a decade from Hiroshima and Nagasaki and only two and a half years from the hydrogen bomb, physicists were viewed as gods who could do anything. Although I did not accept the Glenn L. Martin offer I did profit from the research-grant environment. It was the Air Force who supported my research on quantum gravity as well as postdocs such as Peter Higgs, Heinz Pagels and Ryoyu Utiyama whom I had invited.

But by 1966 the military realized that they weren't going to get magical results from gravity research, and my Air Force grant was terminated. This meant that I was unable to pay the page charges that the *Physical Review* was levying in those days, with the result that the paper I submitted in 1966 was not processed and published until over a year later. Nowadays, of course, everyone goes on the web and receives instantaneous exposure.

The story of ghosts is not the only feature of early quantum gravity research. In 1949 Peter Bergmann [12] began to look for a quantum analog of the 1938 work of Einstein, Infeld and Hoffmann [13] on the motions of singularities in the gravitational field. Since these motions follow from the gravitational field equations alone, without divergent quantities or such concepts as self-mass appearing at any time, Bergmann reasoned that this might be a way to avoid field-theory divergences altogether. Nowadays we regard Bergmann's vision as rather archaic. But it was a vision that was shared by Dirac, who always viewed particles (e.g. electrons) as somewhat different from fields, and even by Feynman, who obtained his famous propagator by carrying out a complex Laplace transform on a heat kernel obtained from a path integral over explicit particle trajectories, both forwards and backwards in time [14–16].

Feynman's conception at least had the merit of being manifestly covariant. But Bergmann had to approach the Einstein–Infeld–Hoffmann picture from the canonical side, singling out the time for special treatment. Although Bergmann's vision never really got off the ground, intensive work was carried out in those years on canonical quantum gravity, culminating in an equation that bears my name along with that of John Wheeler [17] who was the real driving force. Research on the consequences of this equation continues to this day, stimulated by work of Abhay Ashtekar [18, 19], and some of it is quite elegant. But apart from some apparently important results on so-called "spin foams" [20] I tend to regard the work as misplaced. Although WKB approximations to solutions of the equation may legitimately be used for such purposes as calculating quantum fluctuations in the early universe [21, 22], and although the equation forces physicists to think about a wave function for the whole universe and to confront Everett's many-world view of quantum mechanics [23], the equation, at least in its original form [17], cannot serve as the *definition* of quantum gravity. Aside from the fact that it violates the very spirit of general relativity by singling out spacelike hypersurfaces for special treatment, it can be shown not to be derivable, except approximately, from a functional integral [17]. For me the functional integral must be the starting point.

I cannot leave the Einstein–Infeld–Hoffmann–Bergmann–Dirac–Feynman story without mentioning one difference between particles (specifically fermions) and fields (specifically bosonic fields) that raises an issue of terminology. Despite the fact that mathematicians have found the Dirac operator to be more fundamental than the Laplacian [24] (at least in index theory) fermions, unlike bosons, cannot even be introduced into spacetime unless spacetime satisfies certain Stiefel–Whitney conditions allowing the introduction of a spin or pin bundle [25]. By the two-to-one homomorphism from the spin (or pin) group to the Lorentz group, such a bundle defines a Lorentz-frame bundle. A local trivialization of the Lorentz-frame bundles defines a field of local Lorentz frames. These were invented by Elie Cartan who called them *repères mobiles*. There is a well documented exchange of letters [26] in which Cartan tried in vain to get Einstein to understand the value of using *repères mobiles* [27]. When Pauli and others finally got the point the jargon of a German speaking in-group took over, resulting in such barbarisms as *vielbeine* or *bein* rotations. In my view it is inappropriate to use German jargon for something invented by a Frenchman. As for English jargon, "frame" is at least the correct translation of "repère".

After the rules for ghosts to all orders were settled, progress was rapid. 't Hooft and Veltman invented dimensional regularization [28], which kept gauge invariance intact and allowed Zinn-Justin and others to show in detail the renormalizability of the Yang-Mills field minimally coupled to other renormalizable fields [29]. Zinn-Justin's proof of the gauge invariance of counter terms is applicable to the gravitational field, but there the best one can hope for is a low energy effective theory, obtained by minimal subtraction, order by order. Although efforts [30] to make the gravitational field serve as its own cut-off, in some nonperturbative way, had been undertaken several times in earlier years, none of these efforts panned out.

The modern history of quantum gravity begins with Stephen Hawking and his discovery of the thermal radiation emitted from black holes formed by gravitational collapse [31]. The thermality of the radiation allows one to assign a temperature and an entropy to a black hole [32]. The entropy of a solar-mass black hole turns out to be fantastically large, twenty orders of magnitude larger than the entropy of the sun itself. And the disparity is even greater for larger masses. This suggests that the entropy of a black hole is the maximum entropy that any object with the same size and mass can have, an idea that has spurred many attempts to compute the entropy from first principles by summing over putative internal states of the black hole. Perhaps the most successful of these efforts have been string theory computations for certain extremal black holes [33] and the spin-foam[15] computations for Schwarzschild black holes [20].

In viewing string theory one is struck by how completely the tables have been turned in fifty years. Gravity was once viewed as a kind of innocuous background, certainly irrelevant to Quantum Field Theory. Today gravity plays a central role. Its existence *justifies* string theory! There is a saying in English: "You can't make a silk purse out of a sow's ear." In the early seventies string theory was a sow's ear. Nobody took it seriously as a fundamental theory. Then it was discovered that strings carry massless spin-two modes [34,35]. So, in the early eighties, the picture was turned upside down. String theory suddenly *needed* gravity, as well as a host of other things that may or may not be there. Seen from this point of view string theory is a silk purse. I shall end my talk by mentioning just two things that, from a nonspecialist's point of view, make it look rather pretty.

In 1963 I gave [Walter G.Wesley] a student of mine the problem of computing the cross section for a graviton–graviton scattering in tree approximation, for his Ph.D. thesis [36]. The relevant diagrams are these:

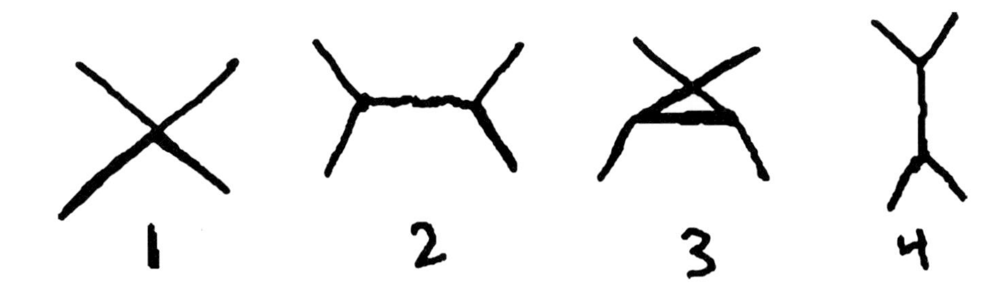

Given the fact that the vertex function in diagram 1 contains over 175 terms and that the vertex functions in the remaining diagrams each contain 11 terms, leading to over 500 terms in all, you can see that this was not a trivial calculation, in the days before computers with algebraic manipulation

[15] A. Corichi suggests that DeWitt meant "spin network".

capacities were available. And yet the final results were ridiculously simple. The cross section for scattering in the center-of-mass frame, of gravitons having opposite helicities, is

$$d\sigma/d\Omega = 4G^2E^2 \cos^{12} \frac{1}{2}\theta \sin^4 \frac{1}{2}\theta$$

where G is the gravity constant and E is the energy [36].

In string theory there is only one diagram, namely and its contribution

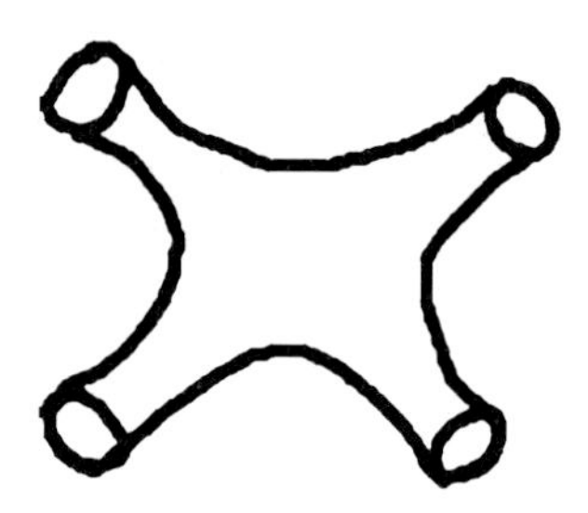

to the graviton–graviton amplitude is relatively easy to compute, giving the same result as that obtained by my student.

The other "pretty" feature of string theory concerns the topological transitions. In conventional quantum gravity topological transitions are impossible. I say this despite occasional efforts that have been made in the past to sum "amplitudes" for different spacetime topologies in "euclidean quantum gravity," "euclidean" being chosen to avoid the singularities necessarily accompanying changes of spatial topology in Lorentzian manifolds. In the first place, euclidean quantum gravity simply does not exist, because the euclidean action is not bounded from below. Moreover, there is no classification of topological transitions analogous to the homotopy classification of paths discovered by Cécile and her student Laidlaw [37], which enables one to assign phases to the contributions to path integrals from different homotopy classes, based on the one-dimensional representation of the fundamental group. Cécile's methods are directly applicable to the Yang-Mills field, for which a precise homotopy classification exists. But no group analogous to π_1 exists for the topological analysis of Lorentzian quantum gravity.

In string theory, on the other hand, one finds that strings can live perfectly well on *orbifolds*, which constitute a certain generalization of manifolds. With orbifolds, even Lorentzian orbifolds [38], topological transitions become possible. Therefore John Wheeler's 40-year-old vision of spacetime foam may be a reality [39]."

References (supplied by C. DeWitt-Morette and B. DiNunno)

1. DeWitt, B.S.: Quantum theory of gravity. I. The canonical theory. Phys. Rev. **160**, 1113–1148 (1967)
2. Rosenfeld, L.: Zur Quantelung der Wellenfelder. Ann. Physik **5**, 113–152 (1930)
3. Rosenfeld, L.: Uber die Gravitationswirkungen des Lichtes. Zeitschrift Fur Physik **65**, 589–599 (1930)
4. DeWitt, B.S.: I. The theory of gravitational interactions. II. The interaction of gravitation with light. Ph.D. Thesis, Harvard (1950)
5. DeWitt, C.M. (eds.): Conference on the role of gravitation in physics. WADCTechnical Report 57–216. (Wright-Patterson Air Force Base, Ohio) (1957)
6. Papers from the Conference on the Role of Gravitation in Physics Held at the University of North Carolina, Chapel Hill, North Carolina, January 18–23, 1957, Rev. Mod. Phys. **29** (1957), 351–546. Introductory Note by Bryce S. DeWitt
7. Feynman, R.P.: Quantum theory of gravitation. Acta Physica Polonica **24**, 697–722 (1963)
8. DeWitt, B.S.: Theory of radiative corrections for non-abelian gauge fields. Phys. Rev. Lett. **12**, 742–746 (1964)
9. DeWitt, B.S.: Quantum Theory of Gravity: (1967) I. The Canonical Theory, Phys. Rev. **160**, 1113–1148. Also reprinted in Quantum Cosmology, Fang and Ruffini, (eds.) (World Scientific, Singapore, 1987). Translated into Russian and published by "Nauka," Moscow (1987)
10. DeWitt, B.S.: Quantum Theory of Gravity: (1967) II. The Manifestly Covariant Theory, Phys. Rev. **162**, 1195–1239. Also reprinted in Gauge Theories of Fundamental Interactions, Mohapatra and Lai, (eds.) (World Scientific, Singapore, 1981). Translated into Russian and published by "Nauka," Moscow (1987)
11. DeWitt, B.S.: Quantum Theory of Gravity: (1967) III. Applications of the Covariant Theory, Phys. Rev. **162**, 1239–1256. Also reprinted in Gauge Theories of Fundamental Interactions, Mohapatra and Lai, eds. (World Scientific, Singapore, 1981)
12. Bergmann, P.G.: Non-linear field theories. Phys. Rev. **75**, 680–685 (1949)
13. Einstein, A., Infeld, L., Hoffmann, B.: Gravitational equations and the problem of motion. Ann. Math. **39**, 65–100 (1938)
14. DeWitt B.S.: Dynamical Theory of Groups and Fields., pp 33 and 31 (eq 6.41, fig 1). (Gordon and Breach Science Publishers, Inc., New York and London), 248 pp. Translated into Russian and published by "Nauka," Moscow, 1987 (1965)
15. Ref. 14 is a reproduction of an article under the same name in Relativity, Groups and Topology, 1963 Les Houches Lectures, Bryce S. DeWitt, Cécile M. DeWitt (eds.), 585–820 (Gordon and Breach Science Publishers, Inc., New York and London, 1964)
16. Feynman introduced his propagator (eq 3) in his article Space-time Approach to Quantum Electrodynamics. Phys. Rev. **76**, 769–789 (1949)
17. DeWitt, B.S.: The Quantum and Gravity: The Wheeler–DeWitt Equation. In: Proceedings of the Eighth Marcel Grossmann Conference, The Hebrew University, Jerusalem, Israel (World Scientific) (1997)
18. Ashtekar, A.: New variables for classical and quantum gravity. Phys. Rev. Lett. **57**, 2244–2247 (1986)
19. Ashtekar, A., Geroch, R.: Quantum theory of gravitation. Rep. Prog. Phys. **37**, 1211–1256 (1974)
20. Ashtekar, A., Baez, J., Corichi, A., Krasnov, K.: Quantum geometry and black hole entropy. Phys. Rev. Lett. **80**, 904. [arXiv:gr-qc/9710007] (1998)
21. Fischler, W., Ratra, B., Susskind, L.: Quantum mechanics of inflation. Nucl. Phys. B **259**, 730–744 (1985)

22. Fischler, W., Ratra, B., Susskind, L.: Erratum. Nucl. Phys. B **268**, 747 (1986)
23. DeWitt, B.S., Graham, N.: The Many-Worlds Interpretation of Quantum Mechanics: A Fundamental Exposition by Hugh Everett III, with papers by Wheeler, J.A., DeWitt, B.S., Cooper L.N., Van Vechten, D., Neil Graham. Princeton University Press, Princeton, NJ (1973)
24. Atiyah, M.F., Singer, I.M.: The index of elliptic operators III. Ann. Math. **84**, 546–604 (1968)
25. Choquet-Bruhat, Y., DeWitt-Morette, C.: Analysis, Manifolds and Physics. Part II, pp. 134–141. Elsevier Science, Amsterdam (1989)
26. Debever, R. (Ed.): Elie Cartan and Albert Einstein: Letters on Absolute Parallelism 1929–1932. Original text in French/ German, English translation by Leroy, J., Ritter, J. Princeton University Press, Princeton NJ (1979)
27. Cartan, E.: La Méthode du Repère Mobile, la Théorie des Groupes Continus et les Espaces Généralisés. Paris, Hermann et cie (1935)
28. 't Hooft, G., Veltman, M.: One-loop divergences in the theory of gravitation. Annales de l'Institut Henri Poincaré Section A **20**, 69–94 (and references therein) (1974)
29. Zinn-Justin, J.: Renormalization of gauge theories. Lect. Notes Phys. **37**, 2–39 (Springer) (1975)
30. DeWitt, B.S.: Gravity: a universal regulator? Phys. Rev. Lett. **13**, 114–118 (1964)
31. Hawking, S.W.: Black hole explosions? Nature **248**, 30–31 (1974)
32. Bekenstein, J.D.: Black holes and entropy. Phys. Rev. D **7**, 2333–2346 (1973)
33. Strominger, A., Vafa, C.: Microscopic origin of Bekenstein–Hawking entropy. Phys. Lett. B **379**, 99–104 (1996)
34. Scherk, J., Schwarz, J.H.: Dual models and the geometry of space–time. Phys. Lett. B **52**, 347–350 (1974)
35. Yoneya, T.: Connection of dual models to electrodynamics and gravidynamics. Prog. Theor. Phys. **51**, 1907–1920 (1974)
36. Wesley, W.G.: Quantum Falling Charges. General Relativity and Gravitation. **2**, 235–245. Ph.D. thesis available from ProQuest 789 E. Eisenhower Parkway P.O. Box 1346. Ann Arbor, MI 48106-1346 (1971)
37. DeWitt, C., Laidlaw, M.G.G.: Feynman functional integrals for systems of indistinguishable particles. Phys. Rev. D **3**, 1375–1378 (1971)
38. Polchinski, J.G.: String Theory Volume I: An Introduction to the Bosonic String and String Theory Volume II: Superstring Theory and Beyond. Cambridge University Press, London (1998)
39. Wheeler, J.A.: Geons. Phys. Rev. **97**, 511–536 (1955)

Progress in Quantum Physics Since the Late Forties

1 Functional Integration as a Major Technique in Quantum Physics

What is functional integration? Answer: It is a theory of integration over function spaces. Functional integrals were introduced in quantum physics[1] by Richard P. Feynman[2] in his Ph.D. dissertation "The Principle of Least Action in Quantum Mechanics" (Princeton, 1942).

Feynman's functional integral (*a.k.a.* path integral and sum over histories) was not accepted readily by physicists. On the other hand, a by-product of Feynman's functional integrals, the powerful Feynman diagrams[3], quickly became famous and widely used by physicists; nevertheless, the integral itself was hardly ever used.

As Barry Simon wrote in the preface to his book on Functional Integration[4], "It seemed that path integrals were an extremely powerful tool used as a kind of secret weapon by a small group of mathematical physicists."

[1] Retrospectively, the path leading to Feynman's integrals can be found in Paul Dirac's book *The Principles of Quantum Mechanics*, 3rd ed. Clarendon Press, Oxford (1947) eq. 60 in §32, *see* "Paul Dirac – A Man Apart" by Graham Farmelo in *Physics Today* Nov. 2009, p. 50 for Dirac as a precursor of current research.

[2] Feynman expressed the probability amplitude of a point-to-point transition as a functional integral. This connection between the operator formalism of quantum physics and functional analysis is powerful.

[3] For an excellent survey of Feynman diagrams from their inception to current research, *see The Institute Letter*, Spring 2009 (Institute for Advanced Study, Princeton NJ).

[4] Barry Simon *Functional Integration and Quantum Physics.* Academic Press, New York (1979).

C. DeWitt-Morette, *The Pursuit of Quantum Gravity: Memoirs of Bryce DeWitt from 1946 to 2004*, DOI 10.1007/978-3-642-14270-3_2, © Springer-Verlag Berlin Heidelberg 2011

Julian Schwinger was not one of them. In a letter to Sam Schweber[5] DeWitt wrote: "The functional integral itself is such a formal object that I've always felt it really would have appealed to Schwinger. It was pure formalism and there are things you can do when you have the functional integral in hand that you cannot easily do in any other way. Because he did not use the functional integral, Schwinger did not discover ghosts[6]. I'm sure he would have, if he felt that he could accept the functional integral."

Nowadays, functional integrals are not only accepted but according to DeWitt [BD 103, p. 173], "often regarded as the quantization rule, superseding all the heuristic rules given in the [first] chapters" of his book *The Global Approach to Quantum Field Theory*. Pages 173–177 of this book give an excellent presentation of DeWitt's conception and use of functional integration.

Feynman integrals were not accepted by mathematicians either. Feynman integrals had been dictated by physics, and it did not fit the then accepted theory of integration. Feynman's method gave the correct conclusions but the rigorous theory of integration forbade the use of the method. In his well known 1972 Gibbs lecture[7] Freeman J. Dyson challenged mathematicians "to construct a conceptual scheme which will legalize the use of Feynman sums."

Solid progress has been made in the mathematical formulation of functional integrals.[8] But DeWitt's construction and calculation of functional integrals is dictated by physics.[9] He constructs a basic functional integral of Quantum Field Theory from Schwinger's variational principle, the principle being derived (heuristically) from the Peierls bracket. The Peierls bracket itself follows from an analysis of measurement in quantum physics.

Bryce DeWitt's major steps in the construction of a functional integral adapted to the varied problems in Quantum Field Theory are as follows (all references and page numbers are to [BD 103], unless stated otherwise) and the following couple of pages may be helpful for reading the original work.

---

[5] An 8-page letter to Sam Schweber dated December 6, 1988; Schweber had asked DeWitt to write his recollections of Schwinger and his memories of the early days at Harvard. The letter is available in the archives stored at the Center for American History (*see* Sect. V.4.).

[6] *See* Quantum Gauge Fields in Sect. I.2.2.

[7] F. J. Dyson "Missed Opportunities", *Bulletin of the American Mathematical Society* **78**, 635–652 (1972); *see* p. 647.

[8] P. Cartier and C. DeWitt-Morette, *Functional Integration, Action and Symmetries.* Cambridge University Press, Cambridge (2006).

[9] A closer collaboration between the physics requirements and mathematics potential is beneficial. As Graeme Segal said recently, "Path integrals capture the essence of Quantum Field Theory and may help formulate a Q. F. T. axiomatic."

- Assume that the transition amplitude $\langle \text{out}|\text{in}\rangle$ can be expressed as a functional Fourier integral [eqs. 10.43, 10.44]

$$\langle \text{out}|\text{in}\rangle_J = \int X[\Phi]e^{iJ\Phi}[d\Phi] \tag{1}$$

$$[d\Phi] = \prod_i d\Phi^i \tag{2}$$

where X is determined by imposing the Schwinger variational principle [eq. (3) below] and is found to be equal to the right hand side of eq. (11). Here the Φ^i are (supernumber-valued) variables of integration in the condensed notation where the generic index i does double duty as a discrete label for the field components and as a continuous label for the points of spacetime [pp. 9–11]. The J_i are external sources [pp. 165–168]; if they suffer variations δJ_i then

$$\delta \langle \text{out}|\text{in}\rangle_J = i \left\langle \text{out}|\delta J_j \mathbf{\Phi}^j|\text{in}\right\rangle \tag{3}$$

Here the $\mathbf{\Phi}^j$ are operators acting on $|\text{in}\rangle$; it follows that

$$\frac{\vec{\delta}}{i\delta J_j}\langle \text{out}|\text{in}\rangle = (-1)^{jF}\left\langle \text{out}|\mathbf{\Phi}^j|\text{in}\right\rangle \tag{4}$$

where F and j are the Grassmann parity (0 or 1) of $|\text{out}\rangle$ and $\mathbf{\Phi}$ respectively. Successive functional differentiations introduce the chronological ordering operator T

$$\frac{\vec{\delta}}{i\delta J_{i_n}}\cdots\frac{\vec{\delta}}{i\delta J_n}\langle \text{out}|\text{in}\rangle = (-1)^{i_1+\ldots+i_n F}\left\langle \text{out}|T(\mathbf{\Phi}^{i_1}\ldots\mathbf{\Phi}^{i_n})|\text{in}\right\rangle \tag{5}$$

- *The Peierls bracket*
The essence of quantum physics in the operator formalism follows from the theory of measurement. It can be encoded in the Peierls bracket [pp. 49–54]. The bracket invented by Peierls[10] in 1952 is a beautiful covariant replacement of the canonical Poisson bracket, or its generalizations, used in canonical quantization. Let A and B be any two physical observables. Their Peierls bracket is defined to be [eq. 4.16]

$$(A, B) := D_A^- B - (-1)^{\bar{A}\bar{B}} D_B^- A \tag{6}$$

Where $\bar{A}$, $\bar{B}$ are the parities of A and B (*see* Sect. III.9). Colloquially, $D_A^- B$ may be called the retarded effect of A on B, and $D_A^+ B$ the advanced effect of A on B. The precise definition follows from the theory of measurement in quantum physics. The operator quantization rule

[10] R. E. Peierls, "The commutation laws of relativistic field theory," *Proc. Roy. Soc.* (London) A **214**, 143–157 (1952).

associates an operator **A** to an observable A; the (super)commutator [**A**, **B**] is given by the Peierls bracket[11]

$$[\mathbf{A}, \mathbf{B}] = -i\hbar(A, B) + O(\hbar^2) \tag{7}$$

DeWitt uses the heuristic quantization rule [eq. 8.26]

$$[\mathbf{\Phi}^i, \mathbf{\Phi}^j] = -i(\mathbf{\Phi}^i, \mathbf{\Phi}^j) \qquad \hbar = 1 \tag{8}$$

for the field operators $\mathbf{\Phi}^i$.

- *The Schwinger variational principle*
  Let $|A\rangle$ be an eigenvector with the eigenvalue A of the operator **A** on a Hilbert space or a Fock space. Let **A** and **B** be two operators of a given system, with A constructed from field operators $\mathbf{\Phi}$ taken from a region of spacetime that lies to the future of the defining domain of **B**. $\langle A|B\rangle$ is the transition probability amplitude from $|A\rangle$ to $|B\rangle$.
  Let **S** be the action functional of a system. Suppose the action functional suffers an infinitesimal change $\delta\mathbf{S}$. It follows from the definition of Peierls bracket [pp. 162–165] that

$$\delta \langle A|B\rangle = i \langle A|\delta\mathbf{S}|B\rangle \tag{9}$$

- *The functional integral*
  Imposing the Schwinger variational principle on the functional Fourier transform eq. (1) yields [pp. 171–172]

$$X[\Phi] = Ne^{i\mathrm{S}(\Phi)}\mu(\Phi) \tag{10}$$

  where N is a constant of integration and $\mu(\Phi)[d\Phi]$ plays the role of a measure, but is not a measure in the sense of a Lebesgue measure. DeWitt notes that the *correct* general solution is not eq. (10) but

$$X(\Phi) = \int d\beta \int d\alpha\; N(\beta, \alpha)e^{i\mathrm{S}(\alpha,\beta;\Phi)}\mu(\Phi), \tag{11}$$

  where "α" and "β" stands for sets of parameters associated with the "in" and "out" region of spacetime, respectively.
  Eq. (11) inserted in eq. (1) has served the needs of Quantum Field Theory well – it is still a heuristic formula. On the other hand, one can consider a functional integral as a mathematical object defined by its domain of integration (i.e. a given function space) and obeying simple basic rules of integration (integration by parts, change of variable of integration). Taking advantage of the heuristic results provided by eq. (11) inserted in eq. (1) and the mathematical results in the Cartier and

[11] *see* eq. (1.10) in Ref. 7. The need for a term of order h^2 is rarely mentioned in the heuristic quantization rule. *See*, for instance, eqs. (8.26, 8.27, 10.32) in [BD 103].

DeWitt-Morette book[12], one is well on the way to having a robust theory of functional integration in Quantum Field Theory.

The Geometry of Gauge Fields

.1 Classical Gauge Fields

Gauge fields play a fundamental role in physics, particularly since the appearance in 1954 of the classical article by Chen Ning Yang and Robert L. Mills.[13]

The phrase "gauge transformation" was introduced in 1918 by Hermann Weyl[14] in a geometrically tantalizing, but unphysical attempt to construct a unified theory for the electromagnetic and gravitational fields, the two long-range force fields. For Weyl a gauge transformation was a local change of scale of the unit length. Einstein pointed out that, in Weyl's scheme, atomic spectra would depend upon the world line of the atom. "Weyl became convinced that this theory was not true as a theory of gravitation; but still it was so beautiful that he did not wish to abandon it and so he kept it alive for the sake of its beauty."[15]

Nowadays, with a somewhat different meaning attached to the word "gauge" and the possibility of describing gauge fields in terms of connections on principal fiber bundles, gauge theories remain at the core of the unification schemes proposed for the four known fundamental interactions: gravitation, electromagnetism, and the weak and strong interactions in particle physics.

Connections on principal bundles and their pullbacks

Beginning with a few words on connections, we show that *their pullbacks*[16] are gauge potentials. Principal bundles are a powerful book-keeping method; it is a straightforward formulation of what look like "tricks" in other formulations. Like all powerful tools, it cannot be used casually. The

[12] P. Cartier and C. DeWitt. *Functional Integration, Action and Symmetries.* Cambridge University Press, New York (2006).

[13] C.N. Yang and R.L. Mills "Isotopic Spin Conservation and Generalized Gauge Invariance", *Phys Rev* **95**, 631 (1954).

[14] H. Weyl, *Raum. Zeit. Materie. Vorlesungen über allgemeine Relativitätstheorie.* Springer, Berlin (1919); translated by H. Brose, *Space, Time, and Matter.* Methuen, London (1922).

[15] S. Chandrasekhar, *Physics Today*, July 1979.

[16] The gauge potentials are not connections, they are the pullbacks of connections.

following brief overview does not replace a user manual.[17] It serves only to introduce words used in the principal bundle theory. A principal bundle consists of four entries:

P	the bundle	$p \in$ P (also called a G-bundle)
X	its base space	$x \in$ X (often spacetime labeled M by DeWitt)
Π	a projection of P on X,	Π : P →X, $\Pi^{-1}(x)$ is a fibre of P at x
G	a Lie group called "the structure group" or the "typical fibre."	

Example: *An electromagnetic field* $F = dA$ *over a spacetime* X; the typical fibre is the abelian group G = U(1); a fibre $\Pi^{-1}(x)$ is diffeomorphic to G.

A connection on P leads to a correspondence between any two fibres along a curve C in X. One says that a point p of the fibre over a point x of the curve is "parallel-transported" along the curve by means of this correspondence. We also say that the curve C described by the *parallel transport of p is a horizontal lift* of the curve C.

The following equivalent definition of connection is less intuitive but easier to relate to gauge potentials than the previous one: A connection on P is a 1-form ω on P with values in the Lie algebra G of G.

Working on the bundle is straightforward; however, in our example, the electromagnetic potential **A** and the field strength **F** are defined on the base space not on the bundle.

The potential **A** is the pullback to X of a connection 1-form ω on P.

The field strength **F** is the pullback to X of a curvature 2-form Ω on P.

Definitions of trivializations, sections, pullbacks, right action on P

First cover M by a set of coordinate patches $\{ U_i \}$; a trivialization of the bundle is the choice of (coordinate) mappings

$$\Phi_i : \Pi^{-1}(U_i) \to U_i \times \mathrm{G}$$

A section s_i of $\Pi^{-1}(U_i)$ is defined by the diagram

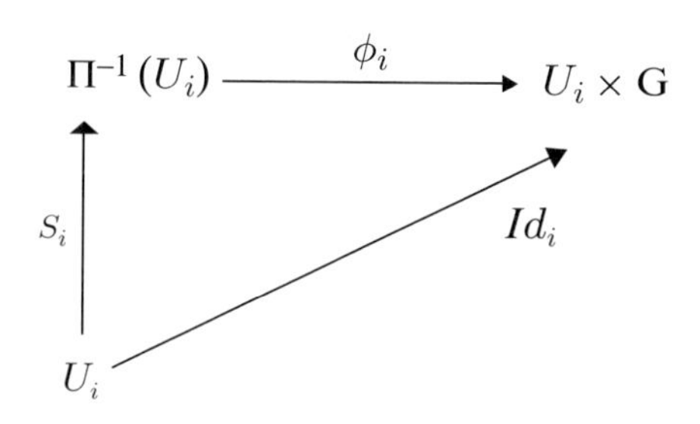

[17] Y. Choquet-Bruhat, C. DeWitt-Morette with M. Dillard-Bleick, *Analysis, Manifolds, and Physics*. North Holland, Amsterdam (pp 357 and 401–410 in the 1989 revised edition). For the correspondence between connection/curvature and the expressions used in physics, *see* pp 364–366, 402–404.

where:

$$Id_i : U_i \to U_i \times \mathrm{G} \text{ by } x \mapsto (x, e)$$

e is the unit element of G. A trivialization Φ_i is usually called a local gauge.

The pullback $\bar{\omega}_i$ of ω is by definition[18]

$$\bar{\omega}_i = s_i^* \omega$$

It is usually called a gauge potential. In the electromagnetic field example above we choose $\bar{\boldsymbol{\omega}}_i = -ie\boldsymbol{A}_i(x)$, where the real function $\boldsymbol{A}_i(x)$ is the electromagnetic potential. If we choose

$$\bar{\boldsymbol{\Omega}}_i = -\frac{1}{2} ie \mathrm{F}_{\mu\nu} dx^\mu \wedge dx^\nu ,$$

the real two form F is the electromagnetic field.

The structure group G has a right action on P, often denoted $\tilde{R}_g$, that does not depend on the choice of trivialization. This global action provides requirements necessary for relating the connections $\bar{\boldsymbol{\omega}}_i$ and $\bar{\boldsymbol{\omega}}_j$ in different local gauges Φ_i and Φ_j corresponding to the same ω on P. The change from $\bar{\boldsymbol{\omega}}_i$ to $\bar{\boldsymbol{\omega}}_j$ is called a gauge transformation.

Electromagnetic theory is a U(1) gauge theory, i.e., an abelian gauge theory. Inspired by the electromagnetic theory, Yang and Mills proposed an SU(2) gauge theory of strong interactions. This is a non-abelian gauge theory.

2.2 Quantum Gauge Fields

Functional integration is a major tool for quantization; the definition of a functional integral rests upon its domain of integration. Therefore quantum gravity requires a study of the domain of integration of its relevant functional integrals, namely a study of the structure and geometry of the space of gauge fields.

DeWitt wrote a review article[19] on the space of gauge fields for the book edited by G. 't Hooft on *Fifty Years of Yang-Mills Theory* [BD 106].

---

[18] Recall:
$s_i'(x)$ maps a vector on U_i to a vector on $\Pi^{-1}(U_i)$
$s_i^*(p)$ maps a 1-form on $\Pi^{-1}(U_i)$ to a 1-form on U_i.

[19] Bryce DeWitt "The space of gauge fields: its structure and its geometry", in: G. 't Hooft (ed.) *50 years of Yang-Mills Theory*. World Scientific Publ, Singapore (2005) pp 15–32. This is the best reference for DeWitt's work on this topic, meaningful to theoretical physicists and mathematicians. DeWitt succumbed to cancer before he could finish this article. The following paragraph is an abstract of the missing sections.

The following quotations are the introduction to this article and the list of references for the article. The references give the list of papers considered important by DeWitt. A complete version of the structure of the space of gauge fields can be found in [BD 103, Chap. 24].

"Introduction

The invention of Yang and Mills [1] was not the first non-Abelian gauge field known to physicists; the gravitational field has that honor. In fact it was the attempt to quantize gravity that yielded, as a by product, the correct rules for quantizing the Yang-Mills field. In 1957 Richard Feynman attended a conference in Chapel Hill [2] at which one of the many questions being argued was whether the gravitational field had to be quantized. Feynman insisted that it did, and he began to think how one would go about it. The results of his thinking [3] were presented at a conference in Warsaw in 1962. Combining two facts, (1) the gauge invariance of tree amplitudes and (2) the expressibility (enforced by unitary) of loop diagrams as sums of products of tree amplitudes, he was able to obtain a gauge invariant and Lorentz covariant expression for the one-loop contribution to any S-matrix element. He found that the non-physical modes carried around the loop by the standard propagator (in any gauge) had to be compensated by another loop involving the propagator for a *fermionic* field, nowadays called a *ghost*.

"In 1964 the author [4] discovered how to handle ghost propagators in two-loop order, using methods that, with enough labor, would yield the correct rules to any order. By the end of 1965 a functional integral was found having the following properties: (1) It produced the one- and two-loop results. (2) It was independent of the choice of gauge-breaking terms[20] in the action and therefore necessarily gave the correct rules to all orders. For a variety of reasons[21] this functional integral did not appear in print until 1967 [5]. Shortly afterwards L. D. Faddeev and V. N. Popov [6] also published the

---

"Use of the geodesic normal field leads to a formalism in which the ghosts disappear and every diagram is individually gauge invariant and independent of the gauge breaking term (59) (*see* Chapter 26 of Ref. [8] [in BD 103]). However, the geodesic normal fields are non-local fields because they depend on Vilkovisky's connection, which is non-local. They also do not have the correct asymptotic behavior to serve as good interpolating fields for the S-matrix. On the other hand, they are excellent for the so-called "closed-time-path" formalism which computes "in-in" expectation values (*see* pp 676 to 684 of Chapter 31 of Ref. [8].) In this case the Vilkovisky connection becomes a "retarded" connection."

[20] often called "gauge-fixing terms".

[21] The author's desire to cast it in the framework of Feynman's sums over products of tree amplitudes, his simultaneous work on canonical quantization, his inability to pay page charges, and delays caused by a referee.

correct rules, in this case derived from a technique that involved integrating over the gauge group."

"In neither of these publications was much attention paid to the structure of the domain over which the functional integrals are taken, i.e. the space of all gauge fields. In subsequent years this omission was partially rectified, particularly when instantons came under discussion, but many aspects remain relatively unknown to this day. It is the purpose of this report to give a clear account of these neglected topics and to note the impact some of them have on conventional ideas about ghosts."

References (list of references as included in the full article [BD 106])

1. C. N. Yang and R. L. Mills, *Phys. Rev.* **96**, 191 (1954).
2. *Conference on the Role of Gravitation in Physics*, Chapel Hill, North Carolina, January (1957) (Wright Air Development Center Document No. AB 118180).
3. R. P. Feynman, *Acta Phys. Pol.* **24**, 697 (1963).
4. B. DeWitt, *Phys. Rev. Lett.* **26**, 742 (1964).
5. B. DeWitt, *Phys. Rev.* **162**, 1195 (1967).
6. L. D. Faddeev and V. N. Popov, *Phys. Lett. B* **25**, 29 (1967).
7. B. DeWitt, *Dynamical Theory of Groups and Fields* (Gordon and Breach, 1965).
8. B. DeWitt, *The Global Approach to Quantum Field Theory* (Oxford, 2003).
9. V. N. Gribov, *Nucl. Phys. B* **139**, 1 (1978).
10. I. M. Singer, *Comm. Math. Phys.* **60**, 7 (1978).
11. G. A. Vilkovisky, in *Quantum Theory of Gravity*, S. M. Christensen ed. (Adam Hilger, Bristol, 1984).
12. I. A. Batalin and G. A. Vilkovisky, *Phys. Lett. B* **102**, 27 (1981); *Nucl. Phys. B* **234**, 106 (1984); *J. Math. Phys.* **26**, 172 (1985).
13. J. Zinn-Justin, in *Trends in Elementary Particle Theory – International Summer Institute on Theoretical Physics in Bonn* (Springer-Verlag, Berlin, 1975).
14. C. B. Becchi, A. Rouet and R. Stora, *Comm. Math. Phys.* **42**, 127 (1975); in Renormalization Theory, G. Velo and A. J. Wightman eds. (Reidel, 1976); *Ann. Phys.* **98**, 287 (1976); I. V. Tyutin, Lebedev Institute preprint N39 (1975)."

The article combines concepts from fibre bundle theory and the tribal language of physicists. Physicists force mathematicians into unchartered territories, but few mathematicians understand the words coined by physicists and few physicists make it easy for mathematicians to read their articles. As a modest foot-bridge to help cross over the physics/mathematics language gap in quantum gauge fields, we give a few equivalences between the words used in DeWitt's article and the words used in fibre bundle theory:

Principal bundle $\mathrm{P} := (\mathrm{P}, \mathrm{M}, \Pi, \mathrm{G})$	the space of gauge fields
$\Pi^{-1}(x)$ a fibre over x	orbit
G structure group (a typical fibre)	gauge group
choosing a section $s_i : U_i \to \Pi^{-1}(U_i)$;	choosing a gauge, choosing a gauge breaking term
changing sections	gauge transformation

pullback to X of a connection 1-form	gauge potentials (also called gauge fields; called connection 1-form in [BD 103] p 94)
pullback to X of a curvature 2-form	field strength
trivialization	fibre adapted coordinates

Now for the word "ghost." Ghosts were introduced by Feynman for constructing functional integrals in non-abelian gauge theories that would be invariant under gauge transformations. The word "ghost" seemed appropriate[22] because the ghost field is a fermionic field compensating a bosonic field and corresponding to non-physical modes. Ghosts have been used extensively and considered necessary for a consistent Quantum Field Theory.

In one of the introductory notes[23] to the book he edited in 2005, 't Hooft gave the mathematical identity of ghosts, "The ghosts discovered by Feynman are due to a jacobian factor in the functional integral." The jacobian identity of ghosts can be established from a careful reading of the articles by L. D. Faddeev and V. N. Popov[24]. It is derived in DeWitt's article[25] (eqs. 45 and 60).

"The most important lesson to be learned from [DeWitt's] derivation is that the ghost arises entirely from the fibre-bundle structure of Φ, from the jacobian of the transformation from the fibre adapted coordinates to the conventional local fields ϕ^i. To get the ghost it is not necessary to integrate over the gauge group[26], which is a dubious procedure at best when the gauge group is the diffeomorphism group of gravity theory."

An excellent exercise would be to rewrite DeWitt's article in the simpler language of a functional integral defined on a principal bundle (rather than constructing it on its base space). Indeed, an action functional is not necessary for constructing a functional integral relevant to Quantum Field

[22] Words such as "ghost", "anomalies", etc. are used by physicists when they come across new, unexpected items. Later on they may be mathematically identified but they continue to be called by their birth names for a long time.

[23] 't Hooft "Ghosts for Physicists" in *50 Years of Yang-Mills Theory*, *see* footnote 19.

[24] L. D. Faddeev and V. N. Popov, "Feynman Diagrams for the Yang-Mills Field", *Phys. Lett. B* **25**, 29–30 (1967)

V. N. Popov and L. D. Faddeev "Perturbation theory for gauge-invariant fields", *Kiev Report* No ITF 67–36 (1967) in Russian, translated into English informally by M. Veltman in 1968, translated by D. Gordon and B. W. Lee in 1972. It appears in 't Hooft ed. loc. cit. pp 40–60.

[25] *see* [BD 106].

[26] The Faddeev-Popov procedure includes an integral over the gauge group.

Theory. One can start with dynamical vector fields[27] on the bundle[28]. And thanks to Vilkovisky's connection on the frame bundle on the space of gauge fields, this construction is possible but remains to be done explicitly!

A natural metric γ on the space Φ of gauge fields. The Vilkovisky connection[29]

The DeWitt blueprint for constructing the "remarkable"[30] connections discovered by Vilkovisky consists of the following:

- The space Φ of gauge fields as a principal fiber bundle over spacetime.
- A natural metric[31] γ on Φ, determined by a metric on the typical fiber G.
- A metric **g** on the base space Φ/G, projection of γ defined on Φ.
- A frame bundle with the same base space Φ/G as Φ, labeled $F(\Phi/G)$ and a riemannian connection on $F(\Phi/G)$ defined by **g**.
- The bundle $F(\Phi/G)$ is different from Φ; its typical fibre is not the gauge group G but the subgroup of the group $GL(n)$ that leaves a quadratic form invariant (euclidean or minkowskian).
- The Vilkovisky connection is a connection on Φ, obtained from the riemannian connection on $F(\Phi/G)$.

The properties of the Vilkovisky connection make it a very desirable tool. It is very tempting to use it to define a functional integral by the method developed for construction of functional integrals on frame bundles.

DeWitt mentions the non-locality of the Vilkovisky connection as a drawback[32]. However, a non-local object in a space can be a local one in a bigger space. For instance, a function $f : \mathbb{R} \to \mathbb{R}$ by $x \mapsto y$ is "nonlocal" in $\mathbb{R}$ but f is a point in an infinite dimensional space of functions.

In 2003, DeWitt was awarded the Pomeranchuk Prize "for discovery and development of quantization methods in gauge theories which laid the foundation for understanding the quantum dynamics of gauge fields."

---

[27] Vector field generating a group of transformations.

[28] For the construction of a functional integral based on dynamical vector fields on a bundle, *see*, for instance. P. Cartier and C. DeWitt-Morette, *Functional Integration, Action and Symmetries*. Cambridge University Press, Cambridge (2006) 135–145, for applications to frame bundles *see* pp 139–141.

[29] For many readers, this section can be enjoyed only as a "spectator sport". Sit down and watch the technician work. *see* [BD 103] for a technical presentation of the Vilkovisky connection.

[30] *see* [BD 106].

[31] Given a metric γ on Φ, one can define horizontal vectors at a point $\phi \in \Phi$ as perpendicular, under the metric γ, to the fiber through Φ. In other words one can define a connection on Φ.

[32] *see* [BD 106].

Progress in Gravitation Since the Late Forties

.1 Numerical Relativity

1.1 Classical Relativity

(Contributed by Richard A. Matzner, Brandon S. DiNunno, and Paul Walter.)

In 1952, upon returning from India where he had been a Fulbright Scholar at the Tata Institute for Fundamental Research in Bombay, Bryce DeWitt took a position as senior physicist at Lawrence Livermore National Laboratories[1], a national lab founded by the University of California under the direction of Edward Teller and Ernest Lawrence. Bryce was originally recruited to investigate properties of neutron diffusion, but through a series of events eventually became the in-house expert on numerical methods.

Bryce's office mate at Livermore (probably Richard Stuart) was working on a two dimensional hydrodynamic code and would often discuss the progress of his project with Bryce. In one dimensional computations, one generally employs lagrangian coordinates due to their high accuracy. Similar accuracy was needed in the two dimensional effort, but as Bryce described in a talk he gave at a symposium for Jim Wilson, everyone at the lab was afraid of using lagrangian coordinates because they would inevitably become curvilinear and require the introduction of jacobians. Bryce reported that, as a relativist, he never could understand why people were so worried, "one evening, breaking the rules of the Lab, I decided to work on the problem at home, actually writing things down on paper. I took the

[1] *see* Sect. III.1.

C. DeWitt-Morette, *The Pursuit of Quantum Gravity: Memoirs of Bryce DeWitt from 1946 to 2004*, DOI 10.1007/978-3-642-14270-3_3,

hydrodynamic equations in two dimensions and differenced them … the jacobians simply drop out and you're left with very simple equations."[2]

Bryce took his results to Teller who immediately called a meeting of the numerics group. The meeting, which resulted in Bryce's promotion to in-house expert on numerical methods, ended with Teller imposing an imminent deadline on the group for converting Bryce's method to actual code. The final version of the code, completed a few weeks later – with an incredible amount of input from Bryce – produced very favorable results. Today, however, even higher degrees of accuracy are obtained through the so called Arbitrary Lagrangian-Eulerian (ALE) method[3], which is a hybrid of Eulerian and lagrangian algorithms. Eulerian algorithms fix the computational mesh to a stationary point, and the fluid then flows with respect to the grid. The advantage gained by Eulerian algorithms is that large changes in the fluid flow are handled fairly well, but at the expense of accuracy.

Lagrangian algorithms, generalized to two dimensions by Bryce, fix the computational mesh to the fluid's moving frame and are very accurate in small domains, but become pathological as the computational domain grows. The ALE method attempts to reduce the pathologies encountered in each individual method by intelligently using both Eulerian and lagrangian algorithms.

Bryce remained at Livermore until 1955. He then moved to the University of North Carolina in Chapel Hill.[4] In 1957, during a seminal conference[5] at Chapel Hill[6], Bryce suggested the use of computers to solve Einstein's equations and to study the full nonlinear structure of relativity: in a discussion session, Mme Fourès[7] mentioned that she can solve a system of equations which are linear with respect to the highest derivatives. "For instance, I can solve them by giving the values of the unknown u in my three-dimensional space Σ on a two-dimensional variety S, and I can solve

[2] [BD 69] and Bryce DeWitt "A numerical method for two-dimensional lagrangian hydrodynamics," **UCRL-4250**. University of California Laboratories, Livermore Site, 10 Dec. 1953.

[3] For more information on the ALE method *see*, for instance, J. Donea, A. Huerta, J.Ph. Ponthot, and A. Rodríguez-Ferran, "Arbitrary Lagrangian-Eulerian Methods", in: *The Encyclopedia of Computational Mechanics*, Vol. 1, Wiley, New York (2004) pp 413–437.

[4] *see* Sect. III.3

[5] *see* Sect. III.4.

[6] A full transcript of the conference can be found in: "Conference on the Role of Gravitation in Physics", *WADC Technical Report* **57–216**; ASTIA Document No. AD 118180, C. DeWitt (ed.) Wright Air Development Center (1957).

[7] Yvonne Choquet-Bruhat, *see* Yvonne Fourès p. 25 of the reference given in footnote 6 in this section; *see* also "Sur l'intégration des équations d'Einstein", *Rational Mechanics and Analysis* **5**, 951–966 (1956). Y. Fourès was the first mathematician to decompose Einstein's equations in 3 space – 1 time dimensions in complete generality. The 3+1 formalism is known nowadays as the shift and lapse formalism.

then the Cauchy problem for the system ... " Bryce then asked if the Cauchy problem was understood well enough to be put on an electronic computer for actual calculation. "Do we now know enough about constraints and initial conditions to do this at least for certain symmetrical cases?" to which Misner responded that he did believe that one could provide initial conditions that should evolve to produce gravitational radiation, and that computers could be used for this. DeWitt then pointed out a few of the difficulties associated with computational techniques, noting that "singularities are of course difficult to handle. Secondly, any non-linear hydrodynamic calculations are always done in so-called lagrangian coordinates, so that the mesh points move with the material instead of being fixed in space. Similar problems would arise in applying computers to gravitational radiation since you don't want the radiation to move quickly out of the range of your computer." Many researchers in the field of relativity believe that this conversation effectively planted the seeds which would later blossom into an entirely new subfield of general relativity – numerical relativity.

Bryce notes that, while in North Carolina, "all these hydrodynamical problems were put out of my mind for years. But in 1970 I had begun to think about the gravitational two-body problem. I thought it was a scandal that nobody had ever tackled this problem. The three-body problem had been a great challenge in Newtonian mechanics. The two-body problem was the analogous challenge in general relativity. I then discovered that my time at Livermore hadn't been wasted. All the lore of differencing partial differential equations came back to me, and I guided my student Andrej Čadež on the first colliding-black-hole computation. Then I moved to Texas (in 1972)."[8] Larry Smarr, who had been a student at Stanford when Bryce gave a course on Relativity (Fall 1971)[9], followed Bryce to the University of Texas at Austin and took the lead in pursuits in numerical relativity.[10]

S. G. Hahn and R. W. Lindquist had originally tried to solve the two body problem in 1964.[11] Their work demonstrated the feasibility of using numerical methods to evolve Misner initial data, but their methods were fatally flawed as such notions of black holes and horizons, area theorems, and the entire scope of the problem had not yet been defined.

---

[8] *see* [BD 69]

[9] *see* Sect. II.2.

[10] L. Smarr, "The Contributions of Bryce DeWitt to Classical General Relativity", in: Steven M. Christensen (ed.) *Quantum Theory of Gravity, Essays in Honor of the 60th birthday of Bryce S. DeWitt*. Adam Hilger, Bristol (1985).

[11] S. G. Hahn and R. W. Lindquist "The two body problem in geometrodynamics", *Ann. Phys.* **29**, 304–331 (1964).

Furthermore, the Misner coordinates they used proved disastrous for computer calculations of the far-field region. Bryce pushed the development of computation for black hole physics. One of the early results in this regard was his 1973 paper – written with F. Estabrook, H. Wahlquist, S. Christensen, L. Smarr, and E. Tsiang – on maximally slicing a black hole [BD 47]. The method they developed set the stage for the collision of two black holes by providing a proper way to represent spacetime with a single black hole. The topology of spacetime in this method is comprised of two asymptotically flat sheets connected by an Einstein-Rosen bridge.[12] Their method of defining the constant-time spaces as maximal has the feature that the time coordinate *stops evolving* before any part of the space reaches the singularity at $r = 0$. In the outer parts of the asymptotically flat region, the time coordinate evolves as it would in flat Minkowski space. Bryce stated[13]: "... it is equally realistic to use the vacuum Einstein-Rosen model for two black holes which replaces each star with a "throat" joining two separate universes ... we will argue below that if the collision of two black holes is capable of radiating gravitational radiation with a high efficiency, then this radiation will be produced during the coalescence of the horizons ... Therefore, we chose to evolve Misner's initial data because they were easiest to work with numerically."

Bryce's foray into computational black hole physics led directly to three Ph.D. dissertations on the subject, two supervised by himself (Čadež, Smarr), and one supervised at Princeton by Smarr (Eppley). The field was kept alive thereafter by Smarr and his students and colleagues. Until the early 1990s, black hole simulations were solely axisymmetric, so black hole collisions were head-on. In the early 1990s, however, numerical relativity was effectively revived in Texas, due to the orchestrated efforts[14] of Richard Matzner and his students and postdocs, who pushed running full 3-dimensional simulations. In 1995 Matt Choptuik joined the faculty and was responsible for attracting a group of students interested in developing adaptive mesh refinement methods later used for the full 3-dimensional simulations.

For at least a decade (from 1995 until 2005) the simulation of merging black holes was plagued by instabilities that became apparent at intermediate times in the evolution, which precluded useful gravitational wave predictions. These were difficult to eliminate because of the very

[12] A. Einstein and N. Rosen "The particle problem in the General Theory of Relativity", *Phys. Rev.* **48**, 73–77 (1935).

[13] *see* [BD 53].

[14] Funded by the NSF grant "The Binary Black Hole Grand Challenge".

complex interaction between the dynamical evolution and the particular gauge chosen. In 2005, Frans Pretorius[15] produced a method using an explicitly hyperbolic formulation that was successful in evolving a merger and the gravitational radiation produced. Subsequently, Baker, Centrella, Choi, and Koppitz[16], and simultaneously Campanelli, Lousto, and Zlochower[17], presented a different method, "moving punctures"[18], which also gave a successful evolution of merging black hole spacetimes. Pretorius's method used a more covariant-like 4-dimensional formulation, while the moving puncture methods are a 3+1 approach to the evolution. The latter is much closer to the formulation of DeWitt and Smarr. The initial data are set as Bryce suggested, but are not *technically* Misner data, rather they are a simpler form intended for nonspinning black holes originally proposed by Brill and Lindquist[19], and later extended by Jim York and others. There is a throat for each black hole in the data; the coordinate system puts the infinity on the other sheet at $r = 0$ where it appears as a singularity. Unlike Misner data which are specifically constructed to have the same geometry on the physical and the "other" asymptotically flat spacetime, here the "other" ends of the wormholes need not even lie in the same manifold. In fact the Misner data have been essentially abandoned because of the effort needed to guarantee symmetry between the two sheets, and because in practice there is no advantage to evolving with this symmetry.

Remarkably the "puncture data" can be evolved, with little contamination of the region near the puncture by the poor resolution of any reasonable coordinates at the puncture. In any case, care has to be taken to ensure that the whole system, including whatever determines the gauges, constitutes a hyperbolic system which is to say that all disturbances should propagate as waves. The groups at Cornell and CalTech have collaborated to produce a spectral implementation of formulations of General Relativity crafted from the beginning to be hyperbolic. Their code, and that of Pretorius, excise the region around the singularity, which logically cannot affect the outside evolution. The Cornell/CalTech code has produced extremely accurate evolutions of merging black holes.

---

[15] Frans Pretorius "Evolution of Binary Black-Hole Spacetimes", *Phys. Rev. Lett.* **95**, 121101 (2005).

[16] J. G. Baker, J. Centrella, D. I. Choi, and M. Koppitz "Gravitational-Wave Extraction from an Inspiraling Configuration of Merging Black Holes", *Phys. Rev. Lett.* **96**, 111102 (2006).

[17] M. Campanelli, C. O. Lousto, and Y. Zlochower, "Last Orbit of Binary Black Holes", *Phys. Rev. D.* **73**, 061501 (2006).

[18] The method allows the black hole singularity to be in the computational domain.

[19] R. W. Lindquist "Initial-Value Problem on Einstein-Rosen Manifolds", *J. Math. Phys.* **4**, 938–950 (1963).

Both Pretorius' and the moving puncture method have been used to produce very interesting simulations including 'whirl' and 'skip' orbits, in which the two black holes come closely together, perform some almost circular orbits and then escape to large distances (the analog of some test particle orbits in the Schwarzschild geometry), and perhaps fall back to repeat the process, losing substantial energy to gravitational radiation on each close encounter until they merge. They have been used to provide a number of predicted waveforms from black hole collisions, and to discover many of the properties of the predicted waveforms, including a kind of dimensional reduction: the waveforms are very simple, even for data sampled in a range of initial configurations (e.g., spin directions, mass ratios, etc.).

The direction of computational relativity now is toward filling out the parameter space, and providing longer (i.e., more physical timescale before and after the merger) and more accurate simulations; and working to provide accurate descriptions including matter, for instance for neutron-star merger modeling. Bryce DeWitt maintained his interest in these simulations until his death, and would be pleased with the strides the field has made since then in particular the interaction between numerical relativity and analytical relativity studies.[20] A valuable contribution of computational physics consists in the exploration of gravitational waves through numerical relativity.[21] The search for gravitational waves with LIGO (Laser Interferometer Gravitational Wave Observatory) will be greatly aided by simulations producing waveforms of binary black hole mergers – the strongest expected emitters of gravitational radiation. By the time the gravitational waves of a merger reach Earth, they are expected to be very weak (wave strain $h \sim 10^{-22}$). In order to pick out a signal from the noise, it becomes necessary to know what to expect. Along with other Numerical Relativity codes, *openGR*[22] will provide expected waveforms for binary black hole mergers. The waveforms from simulations of various initial conditions are combined to construct a template for LIGO for match filtering. Correlating to signals that are stuck in the noise via match filtering will greatly enhance the signal to noise ratio, and thus increase the number of detectable events. *OpenGR* is an open framework for solving problems involving general relativity, namely solving and evolving Einstein's equations, and is available for download at http://wwwrel.ph.utexas.edu/openGR.

---

[20] *see* for instance the plenary lectures of Thibault Damour and Bernd Bruegemann at the 12th Marcel Grossmann meeting (www.icra.it/MG/mg12) on the interaction between numerical and analytical relativity studies.

[21] Paul Walter "Using *OpenGR* for Numerical Relativity", Ph.D. dissertation, University of Texas at Austin (2009).

[22] Developed at the University of Texas at Austin's Center for Relativity.

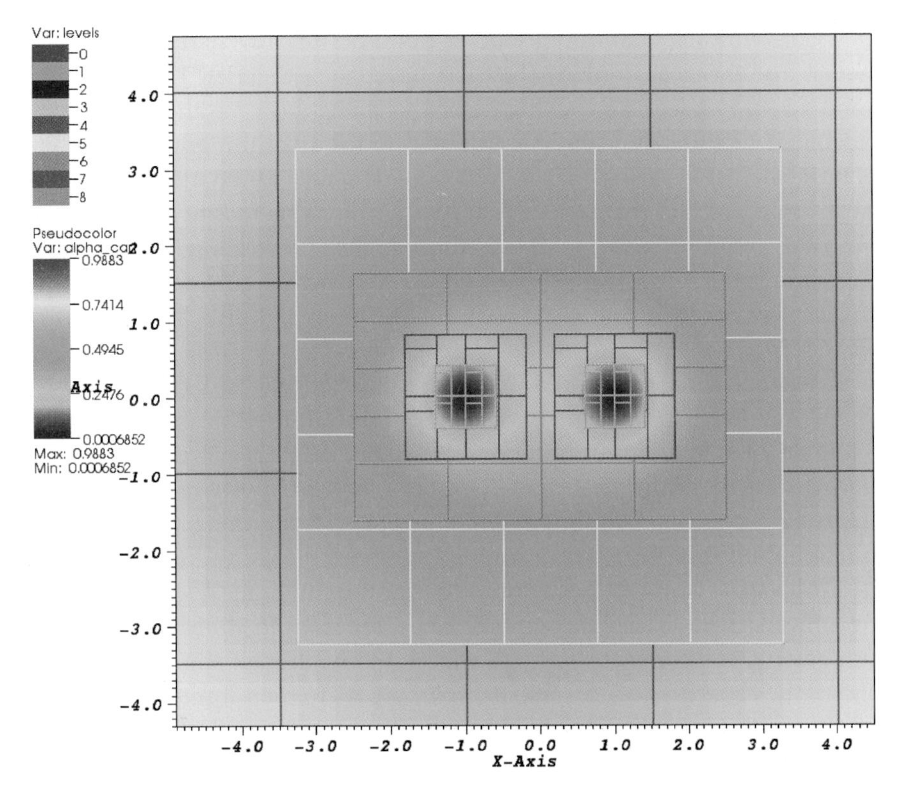

Fig. 2 Zoomed in view of the lapse at the start of a simulation of a head-on collision of two equal mass punctures, or black holes, separated by 2M (where M is the total mass of the black holes). There are nine levels of mesh refinement, with the two finest levels tracking the punctures. The processor layout of the various refinement levels is also shown with the two finest levels tracking the holes. The units on the axes are in terms of the mass of a single puncture (half the total mass $M = M_1 + M_2$). Thus, the punctures are located at $(-\frac{M}{2}, 0, 0)$ and $(\frac{M}{2}, 0, 0)$ and their separation is M

1.2 Toy Models for Quantum Gravity

.2.1 Nonlinear Sigma models in 4 dimensions [BD 81]

The following notes are based on a talk DeWitt gave at the *International Meeting on Geometrical and Algebraic Aspects of Nonlinear Field Theories* held in Amalfi, Italy, May 1988, under the auspices of the Department of Theoretical Physics of the University of Salerno.The proceedings have been edited by S. De Filippo, M. Marinaro, G. Marmo, and G. Vilasi, Elsevier/North Holland, 1989, pp 97–112.

" Conventional quantum gravity is not a perturbatively renormalizable theory. For this reason many theorists have taken the view that it cannot stand on its own feet, and having turned to string theory as the only theory capable of providing an ultimate foundation for quantum gravitational effects. While string theory may indeed provide such a foundation, it is by no means a proven fact that lack of perturbative renormalizability implies that a theory cannot be renormalized at all or cannot be meaningful in its own right in a fundamental sense. No one knows whether conventional quantum gravity exists or not; the question constitutes unfinished business for the theorist.

Repeatedly, in the history of physics, even after interest has turned elsewhere, one has had to come back and settle unfinished business. No way is known of extracting meaning out of a theory that is not perturbatively renormalizable, or, for that matter, out of a perturbatively renormalizable theory in the strong coupling regime, except by computing the Feynman functional integral for a lattice simulation of the theory and attempting to determine an asymptotic behavior in the continuum limit. A small group at the University of Texas[23] has undertaken a program to study lattice quantum gravity. For many reasons this will be a very difficult study, which will stretch the capacities of current supercomputers to their limits and will require some very sophisticated techniques. In order to get their bearings on the subject of lattice simulations of functional integrals and in order to get some preliminary computer experience, the Texas group has decided to look first at a simpler system, which, however, has enough similarities to the gravitational field to make it interesting – namely, the nonlinear sigma model in four dimensions.

There is really an infinity of nonlinear sigma models, each characterized by a group coset space constituting a dynamical configuration space. I shall report here on preliminary results obtained for the $O(2)$, $O(3)$, and $O(1,2)$ models, the configuration spaces being $O(2)$, $O(3)/O(2)$, and $O(1,2)/O(2)$, respectively. It is useful to compare the classical action functional for these models with that of the gravitational field, the latter being given by

$$S = -\mu^2 \frac{1}{2} \int (-g)^{1/2} [g^{\mu\nu}(g^{\sigma\rho} g^{\tau\lambda} - g^{\sigma\tau} g^{\rho\lambda}) g_{\sigma\tau,\mu} g_{\rho\lambda,\nu} \\ + 2 g^{\mu\nu} g^{\sigma\tau} g^{\rho\lambda} (g_{\sigma\tau,\mu} g_{\nu\rho,\lambda} - g_{\mu\sigma,\rho} g_{\nu\lambda,\tau})] d^4x$$

where $g_{\mu\nu}$ is the metric tensor, $g^{\mu\nu}$ its inverse, g its determinant, and μ a scale parameter known as the *Planck mass* (in units for which $h = c = 1$).

[23] Jorge de Lyra, See Kit Foong, Timothy Gallivan, and Bryce DeWitt.

Apart from numerical factors μ^2 is the reciprocal of the gravity constant.

The prototypical form for the classical action of a nonlinear sigma model is

$$S = -\frac{1}{2}\mu^2 \int G_{ab}(\Phi)\Phi_{a,\mu}\Phi_{b,}^{\ \mu}\, d^4x$$

where the fields Φ_a are coordinates in the configurate space; $G_{ab}(\Phi)$ is the group invariant metric on this coset space; commas followed by Greek indices denote differentiation with respect to the spacetime coordinates; the Minkowski metric, which raises and lowers the Greek indices, has signature $-+++$; and μ is a scale parameter having the dimensions of mass."

The results from these calculations can be found in references [BD 81, 83, 85–89].

The following summary of the results has been prepared with the help of See Kit Foong.[24] First a word on the continuum limit of lattice models. The lattice is a discretization of spacetime necessary for numerical computations of Feynman functional integrals. The discretization is a function of N, the number of sites along a side of the lattice and a, the spacing between two adjacent lattice points. Let β be the dimensionless parameter:

$$\beta := \mu^2 a^2$$

depending on the scale parameter μ having the dimension of mass and the lattice spacing a. The bare mass μ is not directly observable, the observable mass μ_R is the result of the interactions encoded in the action functional. Set

$$\beta_R := \mu_R^2 a^2$$

Supercomputers using algorithms based on Monte Carlo methods were used for getting β_R as a function of β. The parameters β and N were the only two adjustable parameters for driving lattice models to their continuum limit as N increases to infinity and a decreases to zero. For the model to be called renormalizable, the renormalized (observable) mass μ_R must remain finite as the continuum limit is approached. With μ_R^2 assumed to be finite, β_R as a function of β must tend to zero as a tends to zero.

As reported in [BD 88], β_R approached zero as β approaches 0 in the three models $O(2)$, $O(3)$, and $O(1, 2)$. The two models $O(2)$ and $O(3)$ were, however, discarded because their corresponding renormalized masses did not remain finite in the continuum limit; they diverge.

[24] In a recent message (April 24, 2009) See Kit Foong recalls his excitement when, contrary to expectation, his calculation showed that the $O(1, 2)$ model is non-renormalizable. "... my mind drifted back to Christmas 1988 when I saw the bending of the $\ln(\beta_R)$-versus-$\ln(\beta)$ curve as β decreases." By then, the group had assumed that the simulation had been done for sufficiently small β to conclude that the model was renormalizable.

The $O(1, 2)$ model remained, and raised hopes that it would be an example of a *perturbatively* non-renormalizable model that was *non-perturbatively* renormalizable. However this optimism did not last very long. After obtaining some discouraging results on the $O(1, 2)$ model, Foong became suspicious of the foundation of the work and carried out a simulation of the simpler $O(1, 1)$ model.

Classically, the $O(1, 1)$ model is the same as the free field model: it is simply the free field model rewritten, through a change of field variables, in terms of two fields satisfying a constraint. Nevertheless the simulations leading to their respective $\beta_R(\beta)$ were astonishingly different. As expected the free field model gave $\beta_R = \beta$. The $O(1, 1)$ model gave a $\beta_R(\beta)$ curve bending towards zero as β decreases – so far so good, *but* the curve appeared to hit a stagnation point at $\beta_R \approx 0.01$ even as β decreases to a figure as small as 10^{-5}. These conflicting results for the $O(1, 1)$ model prompted the group to adopt the geodesic lattice action for the $O(1, 2)$ model, to repeat the whole simulation, and to carry out further analytical, though approximate, calculations. The final result was published four years later, and the stagnation point for the $O(1, 2)$ model was $\beta_R \approx 0.078$. The conclusion is that the quantized $O(1, 2)$ model has no continuum limit in 4 dimensions, or, in other words, the renormalized μ_R diverges as a approaches 0 and the $O(1, 2)$ model is non-renormalizable.

II.1.2.2 The $\lambda\Phi^4$ model[25,26]

The $\lambda\Phi^4$ model is another toy-model for nonlinear fields. It is the simplest Quantum Field Theory. Although the model has been well studied in the past, Bryce DeWitt and Richard Matzner sent a proposal to the NSF in 2001 "to use a lattice-model approach to tie up some loose ends in the theory of the $\lambda\Phi^4$ model in four dimensions, and to gain further insight, with the aid of supercomputers computations, into its properties." The NSF was able to get a response from only one reviewer and postponed its funding decision to FY 03. The proposal was not funded. The proposed work may be pursued by See Kit Foong now Associate Professor at the National Institute of Education of the Nanyang Technology University.

Extracts from comments made by the reviewer and the authors' answers are copied below. The full text includes interesting specifics and technical details.

[25] So-called $\lambda\Phi^4$ because it consists of adding a quartic term, $\lambda\Phi^4$, to the lagrangian of a free scalar field Φ.

[26] The proposal is available in the DeWitts' archives in the Center for American History; *see* Sect. V.4.

RATING: Very Good
REVIEW:

What is the intellectual merit of the proposed activity?
The proposal is to test whether or not the $(\Phi^4)_4$ field theory is trivial by a numerical study that takes into account some of the partial results established mathematically. This is a very basic problem and we should aim to understand it completely. These authors know how to do science well and have access to excellent computational resources. I am very positive except for one glaring defect: There is no discussion of the very many previous numerical studies by other authors! We cannot learn anything from another numerical study unless it includes proper discussion of relative merits and demerits of previous work. I would certainly kill this proposal on this point if I were not impressed with the earlier work of these authors.

What are the broader impacts of the proposed activity?
The $(\Phi^4)_4$ Field Theory is the simplest Quantum Field Theory. It has a renormalizable perturbation theory which predicts non trivial scattering and yet it is believed to be trivial. The same is believed to be true of QED which is the most successful of all Quantum Field Theories. What could be more important than understanding this issue properly?

SUMMARY STATEMENT:
This is an important problem, the authors have experience, high standards, talent and access to the necessary resources. It is good training for junior participants. It should be supported.

Answer to the reviewer's comments

"There is no discussion of the very many previous numerical studies by other authors for the simple reason that we have as yet no numerical results to compare with them. As has been stated at the beginning, our approach will be relatively unsophisticated and will be based on the effective action. *We will not allow ourselves to become sophisticated prematurely.*

This resolve follows from experiences encountered during research on the $O(1; 2)$ nonlinear sigma model (reference 8). In that effort it was felt to be a good idea to begin by studying earlier numerical work on nonlinear sigma models. This proved to be a *bad* idea, which ultimately delayed the getting of solid results by nearly two years.

There are too many red herrings in the literature. We will *not* try to design our numerics to take them into account, at least initially. When we begin to get results we can make comparisons. Merits and demerits of various approaches can then be assessed, and, if necessary, we shall reprogram.

The reviewer, who is clearly interested in the proposal, has kindly made some helpful suggestions."

II.2 Bryce DeWitt's Lectures on Relativity

By 1970, the DeWitt's had begun to think of leaving Chapel Hill.[27] The following drawing by Dan Kennedy summarises the DeWitt's transhumance in 1971 from Chapel Hill to Austin.

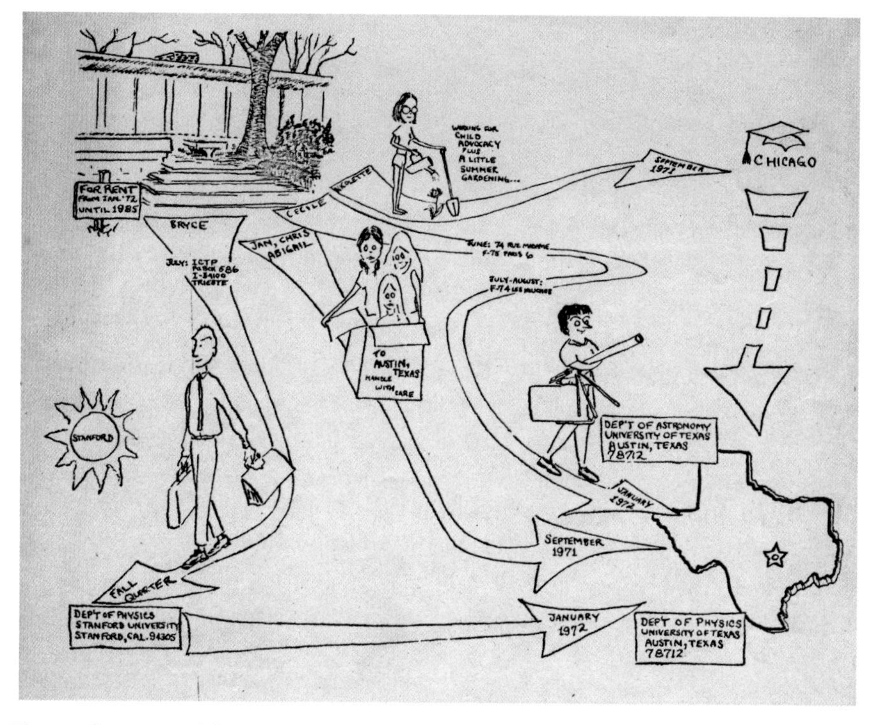

Fig. 3 The DeWitts' "Transhumance" – By Dan Kennedy (1971)

- In July 1971, Bryce went to the International Center for Theoretical Physics in Trieste. He spent the Fall quarter at Stanford University where he gave a course of lectures on gravitation. This section is a report of the publication of his handwritten lecture notes, edited by Steve Christensen. Bryce reached Austin in January 1972.
- Cécile spent July and August 1971 directing, as usual, the Ecole d'Eté de Physique Théorique at Les Houches. She was back to Chapel Hill for the

[27] The reason for the move to the University of Texas at Austin can be read in the biographical memoir written by Steven Weinberg for the National Academy of Sciences; it is reproduced in full in Sect. V.1. The memoir incorporates materials provided to Weinberg by Bryce before his death.

Fall. She joined the University of Texas at Austin Astronomy Department in January 1972.

- Nicolette spent the 1971 Summer in Chapel Hill and went to the University of Chicago in the Fall.
- Chris stayed with Cécile in Chapel Hill during the Fall.
- Jan and Abigail went to Austin in September 1971 ahead of the family to establish a bridgehead for the DeWitts' new home in a new environment. Dan Kennedy's drawing shows, erroneously, Chris going to Austin in the Fall with Jan and Abigail. Dan was a friend of Nicolette and, not paying much attention to Nicolette's young siblings, he shows them in a cardboard box marked "Handle with care."

Bryce DeWitt left a detailed set of handwritten lecture notes for the course he gave at Stanford[28] on gravitation.[29] It contains calculations which cannot be found anywhere else. For instance, Kip Thorne recalls a calcula-

Fig. 4 The four daughters – from left to right: Nicolette, Jan, Chris and Abigail (1990's)

[28] *see* also Sect. IV.1. Larry Smarr, now Director of the California Institute for Telecommunications and Information Technology at the University of California at San Diego, was a student at Stanford University in 1971. When Smarr's advisor Leonard Schiff suddenly died, Smarr, who had attended all of DeWitt's lectures, left Stanford to join the University of Texas at Austin where DeWitt could be his Ph.D. supervisor.

tion he found in the Stanford notes: "When Charles Misner, John Wheeler, and I were writing our book *Gravitation*, I hoped to include a discussion of the flux of angular momentum carried by a source's gravitational waves, based on Richard Isaacson's averaging methods, but I couldn't get the details to work out right. Several years later, when writing an archival paper on "Multipole expansions of gravitational radiation" [*Rev. Mod. Phys.* **54**, 299 (1980), Sect. IV.D], I tried again. I was still flummoxed, so I turned to Bryce for help. He pointed me to his Stanford lectures, where the full details were worked out beautifully. Bryce's mathematical-physics talents were prodigious!"

There were many requests for making these lecture notes available in a book. But who could undertake the enormous job of typing 378 handwritten pages in small writing, mostly equations, minimum wording? Steven Christensen, a former student of DeWitt, volunteered to edit the lecture notes but quickly realized that his own business did not leave him enough time for typing the whole manuscript. In 2005, Luis Alvarez-Gaumé, who had given the Einstein-Prize lecture on behalf of Bryce, approached Christian Caron, Publishing Editor at Springer. Caron, and the editorial board of the 'Lecture Notes in Physics', were immediately enthused by the project and the demanding work on typesetting and editing the notes was the undertaken and eventually completed with the help of Stephen Lyle, who did a magnificent job. They will be published in the same year as this book, as a volume in the series 'Lecture Notes in Physics'.[30]

[29] For more recent works, Cécile DeWitt recommends in particular:

- Yvonne Choquet-Bruhat, *General Relativity and the Einstein Equations.* Oxford University Press, Oxford (2009).
- Valeri P. Frolov and Igor D. Novikov, *Black Hole Physics, Basic Concepts and New Developments.* Kluwer Academic, Dordrecht (1998).

[30] B DeWitt and S Christensen (Ed.) *Bryce DeWitt's Lectures on Gravitation*, Lect. Notes Phys. **826**, Springer-Verlag Berlin Heidelberg (2011).

having opposite sides of equal length. The relationship of such an object to an arbitrary coordinate mesh is indicated schematically in the following figure:

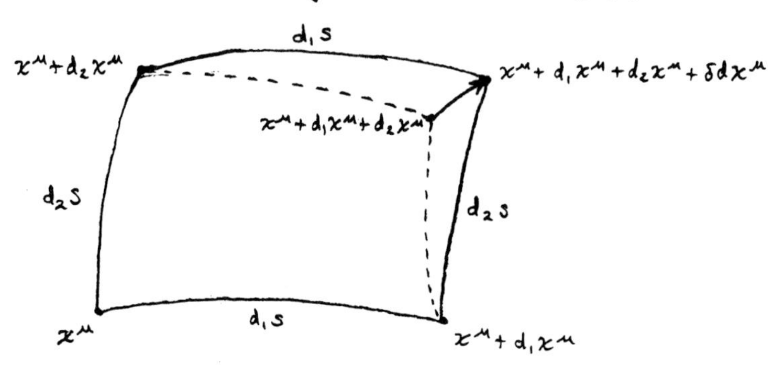

The quantity δdx^μ is the change, arising from the variability in the shape of the coordinate mesh from point to point as well as from changes in the intrinsic geometry of the manifold, in the numerical magnitude of the μ component of <u>either</u> of the infinitesimal intervals d_1x or d_2x as it is displaced in a parallel fashion along the other. It is evident from the figure that

$$d_1s^2 = (g_{\mu\nu} + \tfrac{1}{2} g_{\mu\nu,\sigma}\, d_1x^\sigma)\, d_1x^\mu d_1x^\nu$$

$$= [g_{\mu\nu} + g_{\mu\nu,\sigma}(d_2x^\sigma + \tfrac{1}{2} d_1x^\sigma)](d_1x^\mu + \delta dx^\mu)(d_1x^\nu + \delta dx^\nu)$$

correct to the third infinitesimal order. Keeping terms only up to this order, we find

$$2 g_{\mu\nu}\, d_1x^\mu\, \delta dx^\nu + g_{\mu\nu,\sigma}\, d_1x^\mu d_1x^\nu d_2x^\sigma = 0$$

and, similarly,

$$2 g_{\mu\nu}\, d_2x^\mu\, \delta dx^\nu + g_{\mu\nu,\sigma}\, d_2x^\mu d_2x^\nu d_1x^\sigma = 0 .$$

Two pages from the original handwritten manuscript

$$
\boxed{
\begin{array}{lll}
n_1{}^0 = \sinh a\sigma & n_2{}^0 = 0 & n_3{}^0 = 0 \\
n_1{}^1 = \cosh a\sigma & n_2{}^1 = 0 & n_3{}^1 = 0 \\
n_1{}^2 = 0 & n_2{}^2 = 1 & n_3{}^2 = 0 \\
n_1{}^3 = 0 & n_2{}^3 = 0 & n_3{}^3 = 1
\end{array}
}
$$

$$a_{01} = n_1 \cdot \dot{u}_0 = a \quad , \quad a_{02} = n_2 \cdot \dot{u}_0 = 0 \, , \quad a_{03} = n_3 \cdot \dot{u}_0 = 0$$

$$\dot{\sigma} = (1 + a\xi^1)^{-1} \qquad \boxed{\sigma = \frac{\tau}{1 + a\xi^1}}$$

$$
\boxed{
\begin{array}{l}
x^0(\xi,\tau) = \frac{1}{a}\sinh a\sigma + \xi^1 \sinh a\sigma = \frac{1 + a\xi^1}{a} \sinh \frac{a\tau}{1 + a\xi^1} \\
x^1(\xi,\tau) = \frac{1}{a}\cosh a\sigma + \xi^1 \cosh a\sigma = \frac{1 + a\xi^1}{a} \cosh \frac{a\tau}{1 + a\xi^1} \\
x^2(\xi,\tau) = \xi^2 \\
x^3(\xi,\tau) = \xi^3
\end{array}
}
$$

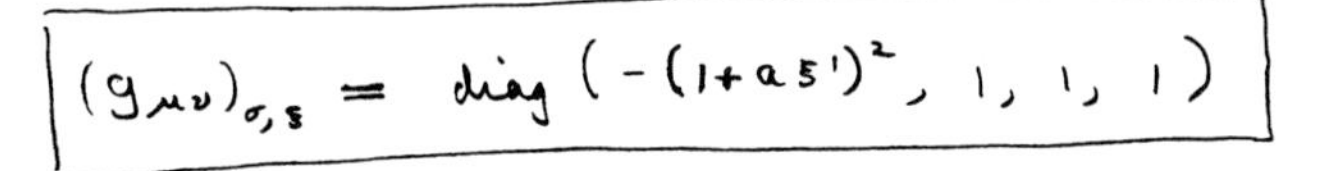

$$(g_{\mu\nu})_{\sigma,\xi} = \mathrm{diag}\left(-(1 + a\xi^1)^2, 1, 1, 1\right)$$

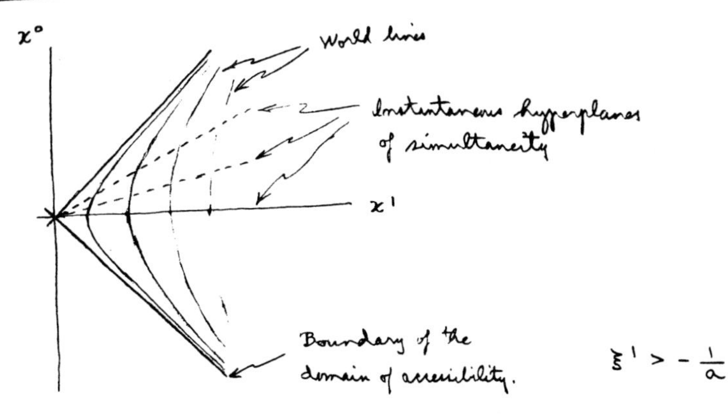

$$\xi^1 > -\frac{1}{a}$$

The book contains 40 problems fully worked out. Most of the problems are conceptual problems, occasionally suggested by science fiction.

In the original lecture notes the equations are not numbered. In the book the numbering has been kept to a strict minimum. As a rule, numbering equations is a disaster, it is used for giving instructions to the reader in connecting equations, rather than giving him/her meaningful sentences.

3 The 1973 Eclipse Expedition[31]

Einstein's theory of gravitation predicts the "bending"[32] of light rays by massive objects. In 1919, Sir Frank Watson Dyson organized two expeditions to measure Einstein's shift during the May 29th eclipse. One expedition led by Sir Arthur Stanley Eddington, Director of the Cambridge Observatory, went to the Island of Principe off the coast of West Africa, the other led by A. C. de la Chiroux Crommelin of the Greenwich Observatory went to Sobral (Brazil).

On November 6, the official results were announced in the main hall of the Royal Society in London. The next morning the general theory of relativity was given front page and editorial space in every important newspaper on the globe. Einstein was suddenly famous.

Dozens of attempts to repeat the British observations were made in the following decades. Of these only six expeditions yielded usable plates; and of these only two yielded results better than marginal. The results of the 1919 expeditions have themselves been repeatedly challenged.[33]

Why are such observations so difficult to make? According to all text books, the observation of the bending of light rays by the sun is explained by the following diagram:

[31] B.S. DeWitt, R.A. Matzner, and A.H. Mikesell "A Relativity Eclipse Experiment Refurbished" *Sky and Telescope* **47**: 301–306 (1974). The original version, available in DeWitts' archives is entitled "Report on the Relativity Experiment at the Solar Eclipse of 30 June 1973".; Texas Mauritanian Eclipse Team "Gravitation deflection of light: solar eclipse of 30 June 1973 I. Description of procedures and final results", *The Astronomical Journal* **81**: 452–454 (1976); D. Evans and K. Winget Harlan's Glove Trotters Xlibris Corporation (2005); Bryce DeWitt "The Story of McDonald Observatory East", (unpublished).

[32] The light ray follows a geodesic. As John Stachel says, "light is going straight in a crooked world".

[33] There is extensive literature on the 1919 eclipse expeditions. The following references provide a good start: S. G. Brush, "Why was Relativity accepted?" *Physics in Perspective* **1**: 84–214 (1999); D. Kennefick, "Testing relativity from the 1919 eclipse", *Physics Today* **62**(3) (March 2009) 37–42.

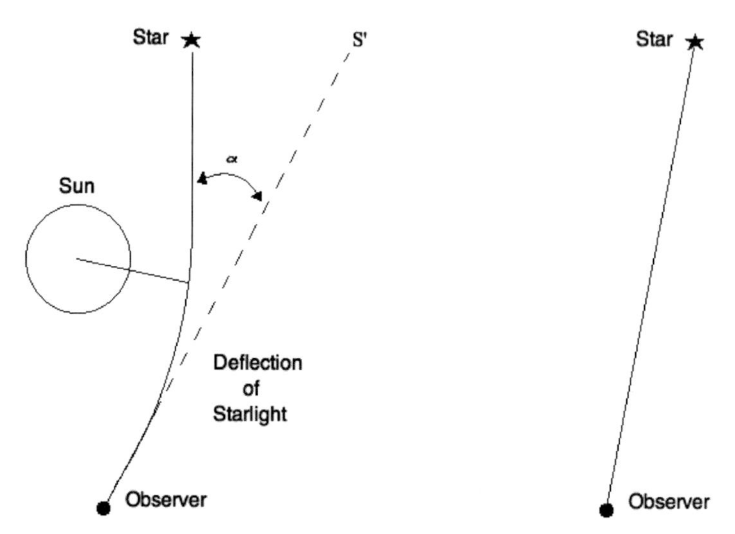

Fig. 5a Daytime, during an eclipse; the observer sees the star at S'

Fig. 5b Nighttime several months later

During an eclipse the stars near the sun are visible and their light rays are bent by the sun (Fig. 5a); several months later the same stars are visible during the night in their normal position. Comparison of the photographic plates taken several months apart give Einstein's shift.[34]

Figure 5a is seriously misleading.[35] Each star image on a plate, when viewed under a microscope, appears as a smudge of darkened grains in a photographic emulsion. Each smudge is at least two thousandths of an inch in diameter (50μ), and the displacement of each from its normal position, due to the bending of light, is less than a tenth of that.

There are many other factors that can cause the smudges to shift in position: creep of the emulsion during the development process, atmospheric refraction, instabilities in the telescope optics due to mechanical stresses or thermal effects, nonuniformity and nonlinearity of the emulsion's response to light. To achieve believable results every one of these factors must be controlled.

Optical observations of Einstein's shift are perhaps the most difficult astronomical measurements ever to be attempted under field conditions. It is not enough to observe a deflection, one has to observe a deflection *different* from the deflection calculated by Johann Georg von Soldner using the Newtonian theory of gravitation as early as 1801.[36]

---

[34] Einstein's shift is not α on Fig. 5a but the angle obtained by superposing Fig. 5b on Fig. 5a.

[35] I did not know any better before I went on the 1973 eclipse expedition and had used Fig. 5a in my classes.

[36] Stanley L. Jaki, "Johann Georg von Soldner and the Gravitational Bending of Light, with an English Translation of His Essay on It Published in 1801," *Foundations of Physics* **9**: 927–950 (1978).

Considering the precision needed to achieve believable results, one can conclude that the shift measured in 1919 was a piece of luck.[37]

The 1973 eclipse is well documented. The following brief writings by DeWitt give a flavor of the expedition.

The final result confirms Einstein's prediction, namely $(0.95\pm0.11)L_E$ where $L_E = 1''75$ is Einstein's value for starlight grazing the sun.

"16 April 1973

To: Sy Alassane, Charles Cobb, Cécile DeWitt, Burton Jones, Richard Mitchell, Richard Matzner

From: Bryce DeWitt

RE: Life and work at Chinguetti

1. This is my first communication to you in my official capacity as director of on-site operations for the eclipse expedition. Most of my communications in the future will, I hope, be verbal, in the field, with lots of feedback from you. It is my fervent desire that my official role be as inconspicuous as possible, and indeed I shall depend on you all in laying out our work program. I think that we shall have little difficulty in deciding collectively our order of work and priorities, but occasionally I may have to make hard or even unpopular decisions. We have a responsibility to The University of Texas, to the National Science Foundation, and to the many other persons, institutions, and agencies that have made this costly expedition possible.
   Therefore I must ask you to conduct yourselves, on site, with quasi-military discipline. I wish I could say that discipline can be forgotten during off-duty hours. But even then we shall have a responsibility to our hosts, the people of Chinguetti, in whose care we shall be leaving our equipment after our departure, until the follow-up team arrives in November.

2. I certainly don't mean to scare you with the word "discipline." The things I shall insist upon will be quite simple and can be largely summed up in the following:

   a. Neatness. I shall expect you to conduct your work and lay out your equipment in a neat and organized way. All tools must be returned to their proper storage niches when you have finished with them, and the darkroom must be left in ship-shape condition.

For light just grazing the sun's edge Einstein's prediction is 1.75 seconds of arc. Newtonian gravitation theory's prediction is just one-half of this, 0.875 second.

[37] The comparison plates used for the Principe plates were taken with different instruments in different locations. The comparison plates used for the Sokral plates were taken with the same instrument in same location.

b. Deliberateness. Careless haste should be avoided at all times. I hope to keep work crises to a minimum, but even when we are under deadline pressure I expect you to work in a deliberate manner. Also, once you have been assigned or have assumed a task for a given work period you should introduce no impromptu innovations in your assignment, or in the manner of carrying it out, without checking first with me.

c. Notebooks. I wish each of you to obtain and keep with you a notebook, or running log, in which you will enter a dated record of your daily work activities. In the case of construction, repair, or maintenance activities a simple note that the job has been accomplished will suffice, but in the case of scientific work I shall expect you to keep a full record of numerical readings, calculations, observations, number and type of exposures made, etc. All exposed plates must be immediately labeled, by means of a code to be decided on site.

d. Water. Until we have ascertained the precise nature of our impact, and that of the other teams, on the water supply of Chinguetti, water is to be used sparingly.

3. I do not yet know precisely what sort of daily schedule will prove to be the most comfortable or convenient for us in Chinguetti. The advance team (Cécile, Burton and myself) will have the opportunity of sizing up the life at the *au gite* and its interaction with the life of the oasis and to come to some consensus in this regard. Tentatively I envisage two possible schedules:

a. 1630 or 1700. Meeting to discuss and plan work for the night. Dinner *au gite.*
1830 or 1900 to 0300 or 0400. Work period.
0400 to 1200. Sleep.
1200 to 1300. Dejeuner au gite.
1300 to 1630. Siesta (or work, if needed).

b. 1800 or 1900. Dinner *au gite.*
1930 or 2000. Meeting to discuss and plan work for the night.
2100 or 2130 to 0600 or 0700 or 0800. Work period.
0700 or 0800. Petit dejeuner *au gite.*
0800–1800. Sleep

Schedule (a) has the advantage of fitting in most easily with the meal hours. Schedule (b) places the work period during the coolest part of the astronomical day and has the advantage of providing some cool daylight hours for construction and maintenance work. In either case arrangements will be made to have food supplied to us during our working hours. I envisage a break at some time in the middle of each work period during which we can relax, eat a bit, discuss the evening's work,

and make any changes in the planned program that may be deemed necessary. Whenever it becomes apparent that we need it, I shall also declare a holiday.
Although we have a tremendous amount to do on site, it is extremely important that none of us become exhausted.

4. There will not be room for more than two (or at most three) persons at one time to sleep in the shelter. Those using it will have to sleep on the floor, but it will be air conditioned, dark during the daytime, and (I hope) relatively quiet. I shall probably set up a schedule for us to take turns sleeping there.

Those of us in Austin are very much looking forward to joining the rest of you in Chinguetti. I think we shall have a great time.

With kindest wishes.
Bryce DeWitt

Copies to H. Smith, D. Evans, A. Mikesell"

12 July 1973
DEPARTMENT OF ASTROPHYSICS
UNIVERSITY OBSERVATORY
SOUTH PARKS ROAD
OXFORD OX1 3RQ
To members of the Texas eclipse expedition:
Charles Cobb, Cécile DeWitt, Burton Jones, Richard Matzner, Richard Mitchell, Sy Alassane

Now that the rigors of the Western Sahara (air conditioned gite, etc.) are over and we have all had a chance to catch up on our sleep and recover a normal digestion, I am venturing to ask how many of you would like to have another go at it? This time we'll take a dozen camels (enough to carry food, bed rolls, and tents) and leave Chinguetti at sundown with a moon four days from full. After exploring the secrets of Richat we shall head southward, across the trackless desert, six hundred miles to Kaedi, remembering always to keep one eye on the lookout for the Chinguetti meteorite.

جامعة تكساس تشكر أهالى شنجيطى للضيافة التى اشتهروا
بها فى مساعدة الذين يأتوا إليهم للبحث عن أسرار الكون

Fig. 6a The following figures show the observatory in Chinguetti. In particular, it also displayed a panel with the following inscription: The University of Texas thanks the people of Chinguetti for the help and hospitality they have always given to those who came to them in search of the secrets of the Universe

Fig. 6b Building the Observatory. Burton Jones sets up the refrigeration by cooling

Fig. 6c The inauguration

Seriously, I want to thank you all most warmly for the wonderful spirit you showed in Chinguetti. I could not have asked for a better crew. Everyone had an indispensable role to fill and he filled it to the utmost of his abilities. It can truly be said: We did the best job possible. In the coming months we shall begin to learn just how successful our effort has been. I wish the final least-squares analysis could reflect the most important parameter of all: your cheerfulness and ungrudging cooperation under tension and stress.

Some of you I shall see again soon. I shall see the rest of you, I hope, before many months have passed. To all of you I extend my kindest wishes.

Bryce DeWitt

The original eclipse plates taken on June 30, 1973 are in the DeWitt's files at the Center for American History at The University of Texas at Austin (*see* Sect. IV.4). A picture of the plates is not included here because no reproduction can demonstrate the fine details needed for the data reduction.

The scientific interest of the expedition is not the only reason DeWitt suggested it to Harlan Smith, then chairman of the Astronomy Department at the University of Texas. DeWitt was always ready to plan or to join an expedition in far away lands:

- On July 4, 1944 he enlisted in the US Navy, and became a naval aviator LtJG with considerable experience in aerobatics.
- In 1950 he went to India as a Fulbright scholar.
- In the following years he rafted down the Omo River from Addis Ababa to the Kenyan border, he went trekking in the high Atlas in Morocco, in Nepal, in Kenya, in the Selous (Tanzania), in the high reaches of K2

Fig. 7 Kip Thorne and Bryce DeWitt, salvaging the scrotum of a bull buffalo that had been shot for food after poachers stole their food supply, on safari in Tanzania, August 18, 1991

coming from the Chinese side, in Ladakh, many climbs in the French Alps and in the Sierras Nevadas.

- His family went along in some of his trips and some of his outdoor activities. He encouraged his daughter Chris to sign up for a parachute class offered to UT students–and down jumped Bryce first, then Cécile, then Chris.

DeWitt kept journals of his trips, recording the events just about daily, even at the end of a long grueling day.

Quantum Gravity

.1 Ph.D. Thesis (Harvard 1950)

Bryce Seligman DeWitt entered Harvard Graduate School in January 1946. He began his thesis work in 1947 under the nominal supervision of Julian Schwinger. The topic DeWitt chose, the quantization of the gravitational field, became his life's work.

DeWitt wrote his recollections of Schwinger at Harvard in response to a request from Sam Schweber.[1] The following is an excerpt from DeWitt's 8-page letter to Schweber.

"You asked me how my thesis subject was determined. It was actually chosen by me. I had studied the book by Bergman on relativity, which included both the special and general theories, and I had also taken the course in relativity that was given at Harvard, although when I took it, it was given by Philip Frank. He was much more interested in the philosophy of relativity than in the detailed mathematics, so I never thought it was a very good course. My interest in relativity therefore, was not really nurtured by anything at Harvard, but was strictly a fascination on my own. I always felt that general relativity was an exceedingly beautiful theory. I also at the same time was appreciating the beauty of some of the things Schwinger was trying to do; and the thought occurred to me, why don't I just go over the lectures on quantum electrodynamics and in place of the electromagnetic field use the gravitational field; try to do the same thing as Schwinger was doing. I thought that I would be led step by step by the lecture notes themselves, and so that it would be a fairly easy thesis. Schwinger agreed to this, but set me an actual task, a goal; in other words although the subject was chosen by me the slant and the aim were set by Schwinger. In (I believe it was) 1930,

[1] Letter to Sam Schweber, December 6, 1988 available at the DeWitt archives, Center for American History, the University of Texas at Austin; *see* Sect. V.4.

C. DeWitt-Morette, *The Pursuit of Quantum Gravity: Memoirs of Bryce DeWitt from 1946 to 2004*, DOI 10.1007/978-3-642-14270-3_4, © Springer-Verlag Berlin Heidelberg 2011

Leon Rosenfeld had calculated the gravitational self-energy of a photon and found that it was quadratically divergent. Schwinger suggested: Why don't we redo this calculation, using his techniques, and show that the notion of an infinite self-energy for a photon is impossible or absurd, and that it, at most, would imply only on an infinite charge renormalization; that is, an infinite rescaling of the electromagnetic field itself.

As long as the calculations were maintained manifestly gauge invariant, it would be impossible for the photon to acquire a mass – which would be the relativistic version of self-energy. Well naturally, since I was copying Schwinger's methods, I didn't attempt anything so crude as a momentum space calculation. Rather, the approach involved what would later become known as current algebra; the selfenergy really involved the commutator of the electromagnetic stress-tensor with itself. Since this was a photon self-energy calculation, rather than a graviton selfenergy calculation, there was no need for ghosts. This was lucky because in those days of course we didn't know anything about ghosts. They weren't invented until much later. They weren't invented until Feynman pointed out the need for them in the early 1960s. Then the perturbation rules for the ghosts to all orders were obtained by me in 1966 and, in a slick, fast technique, by Faddeev and Popov in 1967. In the initial calculations of my thesis I took quantum electrodynamics (that is, the Dirac electron field and the photon field) and added the gravitational field to it. … However, the complications of having three fields were really overwhelming. When I ran into these overwhelming complications I did go see Schwinger. I probably saw him during my work on the thesis a total of about 20 minutes. He told me simply to cut out one of the fields. So I cut out the spinor field[2] and just stuck to pure photons and gravitons. … I shied away from the spinor field because of the complications in needing to use tetrads, or local frames. My thesis itself only included the photon part."

DeWitt made similar comments in the book[3] dedicated to Schwinger's memory, under the title "Remarks before beginning his technical talk" [BD 93a].

A fun memory of graduate school: Archimedian Solids (see Fig. 8) In the winter of 1946–47, Bryce DeWitt and his long-time friend[4], Richard L. Hall, constructed models of the Platonic and Archimedian solids (solid figures whose faces are regular polyhedra) just for fun. From Dick Hall[5], accord-

[2] DeWitt included the spinor field in [BD2].
[3] in: Y. Jack Ng (ed.) *Julian Schwinger, The Physicist, the Teacher, and the Man.* World Scientific, Singapore (1996) pp 29–31.
[4] They were residents of Kirkland house during their college days at Harvard.
[5] E-mail from R. L. Hall to C. DeWitt dated January, 2005. The complete e-mail can be found in the DeWitt archives.

ing to his memory: "It took us about three nights, that is, from about 10 PM to about 3:00 AM, and a half dozen or more file folders, followed by patient measuring, cutting, taping, and some glueing. Fortunately I had taken a mechanical drawing semester in my sophomore year, and the drafting tools helped. Obviously we started with the simple ones first, the tetrahedron, cube, etc. and progressed toward the more complex ones. It occurred to us early on that the more complex ones could be made by cutting off, at an appropriate depth, the vertices of the simpler figures. We finally constructed the most complex figure (what we now know is called the 'great rhombicosidodecahedron'). I am fairly certain we made that one last and that we did the others in more or less logical order of increasing complexity, but I certainly don't recall the sequence exactly. I recall clearly that we put them on the mantel of my room (1-23) and stared at them for a long time, finally concluding that we could see no way to add any more faces."

In his letter to Schweber, DeWitt also recalls a marathon lecture (eleven hours, seven on one day, four the next day) given by Schwinger in 1954 at the Institute for Advanced Study, Princeton N.J. The contents of this lecture were never published but they had a strong impact on DeWitt. Whereas during Schwinger's supervision at Harvard, DeWitt saw him "probably ... a total of about 20 minutes", Schwinger's 1954 lecture motivated him to develop the background field method and to treat simultaneously bosons and fermions.

In 1949, DeWitt went to the Institute for Advanced Study.[6] When Wolfgang Pauli learned that DeWitt was working on the quantization of the gravitational field he "remained silent for several seconds ... and then said 'That is a very important problem. But it will take somebody really smart".[7]

Indeed, Quantum Gravity is to General Relativity what Quantum Electrodynamics is to Special Relativity, and it was Paul Dirac[8] who had worked out the behavior of electrons in a way consistent with both quantum theory and the special theory of relativity. It does indeed take "somebody really smart" to work on quantum gravity.

---

[6] His Ph.D. was granted only in 1950 because his father died during the 1949 summer and he could not complete his thesis before joining the Institute. Until he received his Ph.D. DeWitt was only an unpaid visitor at the Institute.

[7] *Bryce Seligman DeWitt 1923–2004, A Biographical Memoir* by Steven Weinberg, National Academy of Sciences, Washington, D.C. (2008), see Sect. V.I.

[8] G. Farmelo, "Paul Dirac – The Mozart of Science", *The Institute Letter*, Fall 2008, pp 7–9 (Institute for Advanced Study, Princeton N.J) is a brief, illuminating introduction of Dirac.

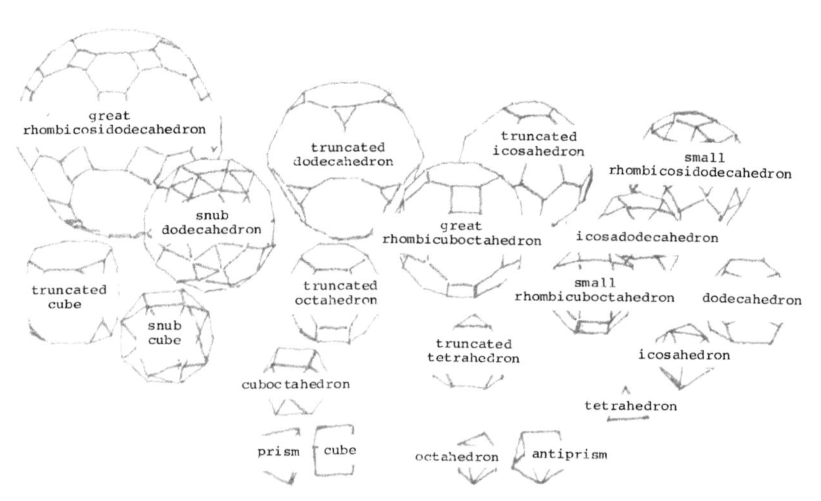

Fig. 8 Archimedian Solids

In 1950, the hamiltonian formulation of Einstein's gravitational field equation had been accomplished by Felix Pirani and Alfred Schild[9], and independently by Peter Bergmann and his coworkers. Pirani and Schild used a hamiltonian formulation, recently developed by Dirac[10], for quantizing the free gravitational field.

In 1951, Bryce DeWitt and the author[11] used Dirac's hamiltonian Dynamics for interacting gravitational and spinor fields. Their paper written while Bryce was a Fulbright Fellow at the Tata Institute for Fundamental Research in Bombay (India) has sometimes been referred to as their "Wedding announcement". The paper rests upon the complicated, and not very illuminating, calculations of the first and second class constraints of the system; but it does develop the formalism of spinor fields in general coordinates. They returned to the United States in 1952 with the first of their four daughters, Nicolette, a ten-day old infant with three birth certificates (USA, French, Indian), two passports, and a middle name Kim (a.k.a. "the little friend of all the world"). Bryce went straight to the US and joined the nuclear weapons laboratory at Livermore, CA. Cécile stayed over in France, with Nicolette, for a few months to run the second session of the Ecole d'Eté de Physique Théorique (Les Houches) that she had founded in 1951. Back in the United States, she joined the faculty of the University of California (Berkeley) as a lecturer.

While in Livermore DeWitt worked on hydrodynamic problems, their formulation and their numerical solutions; the expertise he developed using machine language made it possible for him to launch a numerical relativity program at The University of Texas at Austin in the early seventies (See Sect. II.1) – first for the computations of the behavior of colliding black holes, and then for his students' astrophysical problems.

During his three and a half years at Livermore, DeWitt wrote a treatise on "The Operator Formalism in Quantum Perturbation Theory". In 1953 he won the Gravity Research Foundation prize, then $1,000 (see Sect. III.3).

---

[9] F. A. E. Pirani and A. Schild, "On the Quantization of Einstein's Gravitational Field Equations", *Phys. Rev.* **79**: 986–991 (1950).
F. A. E. Pirani, "On the Quantization of the Gravitational Field of General Relativity", Thesis Carnegie Institute of Technology 1951.

[10] P. A. M. Dirac, "Generalized Hamiltonian Dynamics", *Can. J. Math.* **2**: 129–148 (1950).

[11] Bryce Seligman DeWitt and Cécile Morette DeWitt "The Quantum Theory of Interacting Gravitational and Spinor Fields", *Phys. Rev.* **87**: 116–122 (1952).

III.2 Covariant Quantization *vs.* Canonical Quantization

In brief, the canonical quantization of a system begins with its hamiltonian expressed in terms of coordinates and momenta; its covariant quantization begins with its lagrangian expressed in terms of coordinates and velocities – or rather with its action functional. The hamiltonian formulation of Einstein's equations has been developed by R. Arnowitt, S. Deser, and C.W. Misner[12], using the 3+1 formulations of Yvonne Fourès (*see* footnote 7 in Sect. II.1.).

III.2.1 Dirac's Constrained Hamiltonian Dynamics

By 1949, the path to quantization consisted of well established rules for passing from hamiltonian dynamics to quantum dynamics. But a more general form of hamiltonian dynamics was needed to quantize systems with constraints[13], and Dirac's paper on "Generalized Hamiltonian Dynamics" gave much hope to the handful of physicists investigating the hamiltonian formulation of Einstein's gravitational field equations. This hope can be felt in the introduction to the 1952 paper[14] by Bryce and Cécile DeWitt [BD 2]:

"The hamiltonian formulation of Einstein's gravitational fields which has recently been accomplished by Pirani and Schild[15] and independently

---

[12] R. Arnowitt, S. Deser, and C. W. Misner, "The Dynamics of General Relativity" in: Louis Witten (ed.) *Gravitation – An Introduction to Current Research*. Wiley, New York (1962), pp 227–265; *see also* Yvonne Bruhat, "The Cauchy Problem", pp 130–168 in the same volume.

[13] The first systematic approach to constrained hamiltonian dynamics was published in 1930 by Leon Rosenfeld who applied it to quantum electrodynamics: L. Rosenfeld "Zur Quantelung der Wellenfelder", *Annalen der Physik* **5**: 113–152 (1930); *see also Z. Physik* **65**: 589 (1930). For an excellent article on the early years of Quantum Field Theory, *see* D. Salisbury "Leon Rosenfeld and the challenge of the vanishing momentum in quantum electrodynamics". *Stud. Hist. Philos. Mod. Phys.* **40**: 363–373 (2009).

[14] Bryce Seligman DeWitt and Cécile Morette DeWitt, "The Quantum Theory of Interacting Gravitational and Spinor Fields," *Phys. Rev.* **87**: 116–122 (1952). A number of physicists refer to this paper as the wedding announcement of Bryce, then a Fulbright Fellow in India (1951–1952), and Cécile, then Maitre de Recherches, Centre National de la Recherche Scientifique in Paris, France.

[15] F. A. E. Pirani and A. Schild "On the Quantization of Einstein's Gravitational Field Equation," *Phys. Rev.* **79**: 986–991 (1950).

by Bergmann[16] and his co-workers, has enabled workers in general relativity to consider seriously the possibility of carrying out a rigorous quantization of Einstein's theory. Bergmann and his group hope to develop a quantum theory of the motions of point singularities (particles of matter) in an otherwise "free" gravitational field; i.e., a quantum version of the work of Einstein, Infeld, and Hoffmann.[17] Schild's group, on the other hand, take the more direct course of describing gravitating matter (as well as electromagnetic radiation) by means of additional fields which interact with the gravitational field. The present paper is written in the latter vein."

The paper begins with a treatment of spinors in general coordinates – then it uses Dirac's method for handling constraints, in a form suitable for quantization. The computations of Dirac's first and second class constraints and their mutual coherence are long and tedious. The DeWitts were at that time at the Tata Institute for Fundamental Research (TIFR) then a small quiet Institute housed in a yacht club in Apollo Bunder (Bombay, now Mumbai); it was a good place for working out this difficult problem, conceptually and explicitly.

For an expanded version of Bergmann's work quoted here see [BD 31].

2.2 The Wheeler-DeWitt Equation[18]

The Wheeler-DeWitt equation [eq. 5.5 in BD 31] belongs to canonical quantization although it is very different from conventional canonical quantization. It looks like a Schrödinger equation, but it is a functional partial differential equation, the wave function is a wave function for the universe, the hamiltonian is vanishing.

The genesis of the Wheeler-DeWitt equation is an interesting chapter of the history of Quantum Gravity: DeWitt was known to refer to it as "that damned equation" [BD 95], on the other hand his work on the equation

---

[16] P. G. Bergmann, R. Penfield, R. Schiller, and H. Zatzkis "The hamiltonian of the general theory of relativity with electromagnetic field," Phys. Rev. 80, 81–88 (1950); J. Heller and P. G. Bergmann, *Phys. Rev.* **84**: 665 (1951). For more information on Bergmann's contributions *see* his book *Introduction to the Theory of Relativity.* Prentice Hall, New York, 1946; a paper by Paul Halpern "Peter Bergmann: The Education of a Physicist", *Phys. Perspect.* **7**: 390–403 (2005); and a paper by D. C. Salisbury "Peter Bergmann and the invention of constrained hamiltonian dynamics", (arXiv:physics/0608067, 2006). P. G. Bergmann "Non-linear field theories", *Phys. Rev.* **75**: 680–685 (1949).

[17] A. Einstein, L. Infeld, and B. Hoffmann, *Ann. Math.* **39**(1), 66 (1938); A. Einstein and L. Infeld, *Can. J. Math.* **1**, 209 (1949).

[18] It has also been called the "Einstein-Schrödinger equation".

[BD 31] is his most cited paper.[19] Thanks to the proceedings of the Marcel Grossmann 8 conference held in Jerusalem in June 1997, we have the story in DeWitt's own words [BD 95].

"The Quantum and Gravity: The Wheeler-DeWitt equation

by Bryce DeWitt

Abstract

This equation should be confined to the dustbin of history[20] for the following reasons: 1) By focussing on time slices it violates the very spirit of relativity. 2) Scores of man-years have been wasted by researchers trying to extract from it a natural time parameter. 3) Since good path integral techniques exist for basing Quantum Theory on gauge invariant observables only, it seems a pity to drag in the paraphernalia of constrained hamiltonian systems. 4) In the case of minisuperspace models, gauge invariant transition amplitudes defined by the path integral do not satisfy any local differential equation; they satisfy the Wheeler-DeWitt equation only approximately."

After a few words recalling his invitation to the MG 8 conference, DeWitt explains his position.

"John Wheeler, the *perpetuum mobile* of physicists, called me one day in the early sixties. I was then at the University of North Carolina in Chapel Hill, and he told me that he would be at the Raleigh-Durham airport for two hours between planes. He asked if I could meet him there and spend a while talking quantum gravity. John was pestering everyone at the time with the question: What are the properties of the quantum mechanical state functional Ψ and what is its domain? He had fixed in his mind that the domain must be the space of 3-geometries, and he was seeking a dynamical law for Ψ.

I had recently read a paper by Asher Peres[21] which cast Einstein's theory into Hamilton-Jacobi form, the Hamilton-Jacobi function being a functional of 3-geometries. It was not difficult to follow the path already blazed by Schrödinger and write down a corresponding wave equation. This I showed to Wheeler, as well as an inner product based on the Wronskian for the functional differential wave operator. Wheeler got tremendously excited at this and began to lecture about it on every occasion.

---

[19] Interestingly [BD 31] is the only reference given to the Wheeler-DeWitt equation in Wikipedia. In [BD 31] DeWitt says "The present paper is the direct outcome of conversations with Wheeler.".

[20] The Wheeler-DeWitt equation is not even listed in the index of [BD 103], his magnum opus written at the end of his life.

[21] "A. Peres "On Cauchy's problem in General Relativity – II", *Nuovo Cimento* **26**, 53–62 (1962)".

I wrote a paper on it in 1965, which didn't get published until 1967 because my Air Force grant was terminated, and the Physical Review in those days was holding up publication of papers whose authors couldn't pay the page charges. My heart wasn't really in it because, using a new kind of bracket discovered by Peierls, I had found that I could completely dispense with the cumbersome paraphernalia of constrained hamiltonian systems and build a manifestly gauge covariant quantum theory *ab initio*. But I thought I should at least point out a number of intriguing features of the functional differential equation, to which no one had yet begun to devote much attention: [...] The fact that the wave functional is a *wave function of the Universe* and therefore cannot be understood except within the framework of a many-worlds view of quantum mechanics [....] In the long run one has no option but to let the formalism provide its own interpretation. And in the process of discovering this interpretation one learns that time and probability are both *phenomenological* concepts.

[...] As I told Tsvi Piran, if the organizers of this conference really wanted me to talk about the Wheeler-DeWitt equation they should be quite aware where I stand on it. It has played a useful role in getting physicists to frame important and fundamental questions, but otherwise I think it is a bad equation, for the following reasons: (1) By focusing on time slices (spacelike 3-geometries) it violates the very spirit of relativity. (2) Scores of man-years have been wasted by researchers trying to extract from it a natural time parameter. (3) Since good path-integral techniques exist for basing quantum theory on gauge invariant observables only, it seems a pity to drag in the paraphernalia of constrained hamiltonian systems.

I subscribe 100% to the modern view that the quantum theory should be defined by the path integral. I am going to show you how the path integral can be used both to resolve the conceptual issues and to yield gauge invariant transition amplitudes that are operationally well defined. Except in special cases these amplitudes do not satisfy any local differential equation. They satisfy the Wheeler-DeWitt equation only approximately. This means that, generically, the Wheeler-DeWitt equation is *wrong*, even assuming that the difficult issues of quantum gravity's perturbative nonrenormalizability can be resolved, via string theory or whatever. One may legitimately use the Wheeler-DeWitt equation, and the *WKB* approximations to its solution, in analyzing such things as the role of quantum fluctuations in the early universe.[22] But it is wrong to use it as a definition of quantum gravity or as a basis for refined and detailed analyses.

As I told the conference organizers, decades have passed since I last gave more than a passing glance at the Wheeler-DeWitt equation, and therefore I beg forgiveness of those many persons of whose work I am largely

[22] It has indeed been applied many times to problems in quantum cosmology.

ignorant and will fail to acknowledge. I shall have nothing to say of the important work of the Ashtekar school.[23] I shall also have time only for passing reference to the hoped-for future role of the path integral in the consistent-histories framework for viewing the wave function of the universe, with which I am spiritually in full accord. I hope these lacunae will be filled by others during this conference."

III.2.3 The "Trilogy"

The three following papers, published in 1967, are colloquially called "the trilogy." They were submitted to *Physical Review* as a single paper but were divided into three papers at the request of the Editor:

I. The Canonical Theory [BD 31], his most cited paper.
II. The Manifestly Covariant Theory [BD 32], his 2nd most cited paper.
III. Applications of the Covariant Theory [BD 33], his 4th most cited paper.

DeWitt's covariant quantization brings together the Peierls bracket, Schwinger's variational principle, and Feynman integrals in a unified formalism. That is to say, it consists of an analysis of the disturbances created on a system by quantum measurements (Peierls bracket), the effect of these disturbances on the action functional of the system (Schwinger's principle), and the Feynman integrals that make it possible to compute physical observables of the system.

Papers II and III together with DeWitt's Les Houches lecture notes [BD 23, 29], [BD 65, 66] and his *magnum opus* [BD 103] are the landmarks of his covariant quantization of Quantum Gravity from 1963 to 2004.

Naturally[24] *The Global Approach to Quantum Field Theory* does not deal with canonical quantization; but its preface[25] contains a fitting conclusion for this section on covariant vs. canonical quantization, namely:

"The global approach does not prevent one from appreciating the traditional canonical theory. In appropriate situations, canonical methods are both highly useful and strikingly beautiful. But it is generally easier to descend to them from the global vantage point than to climb in the reverse

[23] Good references to Ashtekar's work can be found in his recent paper "Some surprising implications of background independence in canonical quantum gravity," *General Relativity and Gravitation* **41**: 1927–1943 (2009) [Jürgen Ehlers issue].
[24] A partial quote from DeWitt. The original has not been found. "After having marvelled at the beauty of the covariant formalism ...".
[25] The preface is reproduced in full in Sect. III.10.

direction. They are always accessible and can be brought into play whenever it is convenient to do so. It is often convenient when the specifically (3+1)-dimensional character of spacetime is of primary importance, for example when there exists a global timelike Killing vector field, or when thermal properties are under study.

One cannot do without (3+1)-dimensional assumptions. Although space and time together comprise a single geometrical entity, individually they are distinct."

4.3 The Gravity Research Foundation Essay (1953)

The Gravity Research Foundation was founded by Roger W. Babson in January 1949. It offered awards to be given for suggestions for anti-gravity devices. Needless to say "the scientific community responded with a resounding lack of enthusiasm".[26] There is no professional byline on any of the winning essays from 1949 to 1952, except for the 1950 essay of Richard A. Ferrell, then a graduate student at Princeton University. DeWitt, not limited by *qu'en dira-t-on*, submitted an essay "New Directions for Research in the Theory of Gravitation", refocussing the interest of the foundation from crackpot pursuits to useful endeavors. After he won the prize, the competition was no longer taboo. George Rideout, President of the Foundation, modified the announcement of the essay competition. It is now a highly respected competition that has produced a number of important essays.

"New Directions for Research in the Theory for Gravitation

by Prof. Bryce DeWitt
Radiation Laboratory
University of California

Before anyone can have the audacity to formulate even the most rudimentary plan of attack on the problem of harnessing the force of gravitation, he must understand the nature of his adversary. I take it as almost axiomatic that the phenomenon of gravitation is poorly understood even by the best of minds, and that the last word on it is very far indeed from having been spoken.

Nevertheless, the theoretical investigation of gravitation has received relatively little attention during the last three decades. There are several reasons for this. First, the subject is peculiarly difficult; the existing body of theory

[26] Founding of the Gravity Research Foundation www.gravityresearchfoundation.org.

ESSAYS ON GRAVITY

SIR ISAAC NEWTON

FIVE WINNING ESSAYS
OF THE
ANNUAL AWARD
(1949-1953)

OF THE

GRAVITY RESEARCH FOUNDATION

•

NEW BOSTON
NEW HAMPSHIRE

Founded by Roger W. Babson
in 1948

on it involves rather recondite mathematics, and the fundamental equations are almost hopeless of solution in all but a very few special cases. Although the accepted theory is motivated by two or three beautifully simple yet profound principles, these guiding principles have so far been of little help in predicting the general features of the solutions of the equations to which they give rise. And, as any researcher in the field knows, one can develop a serious case of "writer's cramp" in the manipulation of tensor indices which is usually necessary in order to prove only a single tediously trivial point.

Secondly, modern gravitational theory has few consequences which are even remotely susceptible of experimental verification. The old Newtonian theory, involving action-at-a-distance, has, for practical purposes, been far too adequate. Consequently, stimuli for the theoretical investigation of gravitation are virtually non-existent, and gravitational research is almost totally unrewarding. It is a field which had its brief brilliant hour, but which has since fallen into a state of near disrepute.

In spite of all this, it is very probable that the phenomenon of gravitation will eventually have to be reckoned with again in respectable circles. And it may well happen that this reckoning will present itself in a rather acute form. It is one of the purposes of this note to suggest that we may be already in the first phases of such a new development, and to point out some new directions into which we are likely to be led as a result.

I shall assume, virtually without question, the validity, *in its appropriate domain*, of the Einstein theory of gravitation—that is to say, of the original general theory of relativity, as distinct from later embellishments by many workers including Einstein himself. Einstein's theory is, to my mind, far too beautiful and satisfying to be cast aside. And it is so intimately connected with and firmly intrenched in those concepts of invariance and conservation which have come to be regarded as fundamental in physics, that, in casting it aside, we should be casting aside much that has been enormously fruitful in the past as well as the present, to the experimenter no less than to the theorist. However, it should be borne in mind that the Einstein theory is a "classical" (that is, *non quantum*) theory. It forms by itself a logical and self-contained system. Only the fact that the real world around us has taught us that the system may not be quite so self-contained after all, makes the following remarks of some interest.

For the sake of orientation let us reverse the usual order of things and first fix our sights on those grossly practical things such as "gravity reflectors" or "insulators," or magic "alloys" which can change "gravity" into heat, which one might hope to be the useful byproducts of new discoveries in the theory of gravitation. The use of terms such as "reflector" or "insulator" clearly is based upon analogy with electromagnetism. Now, it is quite true that gravitation is similar to electromagnetism in many ways. Just as the latter can be split into an electric and magnetic part, so can the former be split into two parts, one being that produced by static matter and the other that produced by moving matter. The gauge group of electrodynamics has its counterpart in the *coordinate transformation group* of gravidynamics. The electromagnetic and gravitational fields both propagate with the speed of light.

In other respects, however, the gravitational and electromagnetic fields differ profoundly. Of prime importance is the extreme weakness of gravitational *coupling* between material bodies, as compared with that of electromagnetic coupling (advice of professional weight lifters notwithstanding!). The weakness of this coupling has the consequence that schemes for achieving gravitational insulation, via fanciful methods such as oscillation or conduction, would require masses of planetary magnitude. And even if the necessary masses could be manipulated, these schemes would be doomed to failure, for, since quantum forces would not be available for such macroscopic manipulation, non-gravitational force fields would have to be employed. But the existence of such external fields would defeat its own purposes, because every stress, every force potential, and, indeed, every form of energy produces its own gravitational field. The gravitational field is all-pervading.

These features are built into the Einstein theory as consequences of the fundamental requirements of energy-momentum conservation. One result is that the gravitational field partially produces itself! Mathematically this is reflected in the strong non-linearity of the gravitational field equations, which stands in sharp contrast to the linearity of the electromagnetic field equations.

These considerations are quite sufficient to enable one to state *flatly* that any frontal attack on the problem of harnessing the power of gravity along the above lines is a waste of time. Indeed, unless the term "gravity" is broadened to include a much wider range of phenomena than hitherto, one may safely pronounce all gravity-power schemes impossible. Such a broadening of terminology may, however, be logically possible, or even necessary. That is the point I wish to make.

In one very important respect the Einstein theory has recently undergone some fundamental broadening. It has been "quantized"— or rather, as matters stand at present, "nearly quantized." For a generation, two tree-like giants, the quantum theory, and the general theory of relativity, have existed side by side, one incredibly fruitful and the other almost totally barren save for one or two golden fruits. Except for the most indirect contacts in cosmological problems they have remained completely independent, although the *special* theory of relativity has long since been combined with the quantum theory, with results which, while profound, have not been as successful as one might have hoped. However, in 1950, Pirani and Schild [1] and, independently, Bergman and his co-workers [2] accomplished the *hamiltonian formulation* of the Einstein field equations. By "hamiltonian formulation" is meant a certain *canonical* way of writing the equations which forms the point of departure of a quantum theory of them. With this important accomplishment it became possible for the first time to consider seriously a rigorous quantization of the gravitational field.

When one attempts, however, to pass from the classical hamiltonian formulation of Einstein's equations to a quantum version of his theory, one runs immediately into several problems. Chief among these is the fact that the *hamiltonian density* for the theory contains products involving *non-commuting* factors. One does not know *a priori* how these factors should be ordered.

The quantum hamiltonian density must be an *Hermitian operator*. This implies a symmetrical ordering of the aforementioned factors. One could attempt to use the simplest possible symmetrical ordering scheme, but then one would not know whether an equivalent quantum theory would have been obtained with a similar symmetrization procedure, if another set of variables had been used with which to represent the gravitational field, i.e. if a *point transformation* had been carried out on the field variables. The representation which is most frequently employed is that in which the components $g_{\mu\nu}$ of the metric tensor of space-time are chosen as the gravitational field variables. The present writer has shown [3] how a geometry can be introduced in a natural way into the 10-dimensional "space" of the $g_{\mu\nu}$ (to be distinguished from the 4-dimensional space-time manifold) and has used this geometry to construct an invariant quantization rule. The method is applicable to all systems having hamiltonians which are quadratic in the *momenta*.

$g_{\mu\nu}$-space, according to its natural geometry, is found to be non-flat. This is a strict characterization of the fact that Einstein's theory is intrinsically non-linear. There exists no Cartesian representation in which its quantization can be carried out in simple fashion. It is therefore quite fortunate that an invariant quantization prescription nevertheless exists.

In addition to invariance under point transformations of the gravitational field variables, the quantized theory must also be investigated with respect to the more important question of general *covariance*. The *forms* of all quantum equations must remain invariant under general coordinate transformations. The classical hamiltonian formulation is covariant because it proceeds from a set of covariant equations. After passage is made to the quantum theory, however, the covariance must be proved all over again because 1) the non-commutativity of factors may introduce new difficulties, and 2) the formal appearance of the quantized theory, in the so-called *Schrödinger representation* which one arrives at, is quite different from that of the classical theory. The present writer has examined (unpublished) the anatomy of general coordinate transformations, as seen from the viewpoint of the quantum theory, in considerable detail. The unfortunate result of these researches is the discovery that the quantized theory is no longer covariant.

At this point one may well ask to know the reasons for attempting quantization of the gravitational field in the first place. As a matter of fact, the overwhelming weight of opinion of physicists is opposed to the attempt. The prime reason for this is the experimental fact that gravitation has never been observed to take part in physical events on a quantum level, and where there is no evidence it is bad form to speculate. Even if the covariance failure mentioned above could be regarded as definite negative evidence, it would cause no upheaval in physics. It may actually be that the gravitational field is the one and only field which is *not* quantized in Nature. The gravitational field, with its attendant phenomena, could, under these circumstances, constitute the ultimate classical level which must be postulated, even in the quantum theory, in order to have a consistent "quantum theory of measurement." [4] The gravitational field could be produced, not by the quantum *stress tensor* of all the other fields in Nature, but rather by the quantum *mean value* of this tensor. If ψ denotes the quantum *state vector* of the quantized fields, it would satisfy an invariant Schrödinger equation of the form.

$$i\hbar c \frac{\delta \psi}{\delta x^{\mu}} = H_{\mu} \psi, \text{(the symbol } \hbar = \frac{h}{2\pi}\text{)} \qquad (12)$$

where δx^{μ} is a "time-like" displacement. The hamiltonian density H_{μ} would, of course, depend on the metric tensor $g_{\mu\nu}$ as well as on the quantized field variables. But $g_{\mu\nu}$, depending on the mean stress, would depend on ψ and its adjoint ψ^*. Hence equation (1) would be non-linear in ψ and one of the most fundamental principles of the quantum theory, namely the *principle of superposition of states*, would be invalidated. However, this principle would be invalidated only in the large; it would still be true at the quantum level. The dependence of H_{μ} on the $g_{\mu\nu}$ would be important only on a cosmic level. And here, the superposition principle is of no consequence. The universe is in one and only one state ψ. There exists, so far as we know, no coupling with forces *outside* the universe which could cause a transition of the universe to a different state. The state vector ψ will satisfy equation (1) for all time.

However, even if the gravitational field is left unquantized, there remain difficulties. To mention only one, if an "impressed" unquantized gravitational field is allowed to interact with the vacuum fluctuations of those fields which are quantized, a polarization effect will result which is non-calculable owing to *divergence* difficulties, and which cannot even be handled by modern "renormalization" techniques [5]. This is a situation which is not aggravated by quantization of the gravitational field. In point of fact, at the present stage of the game, there is little to choose between the two possibilities. Improved methods of computation and/or interpretation may overcome the difficulty mentioned here, as well as the covariance-failure of the quantized theory mentioned previously.

There still remains a powerful aesthetic argument on the side of quantization. The dream of a "unified field theory" is as tantalizing as ever. But in a consistent unified field theory one could hardly exempt one part and not the other of a "super field" from quantization. Furthermore, a unified field theory may well be the solution to some of the outstanding fundamental problems of the present day. A unifying principle of some sort is clearly needed to solve the problem of the "mass spectrum" and to bring a semblance of order into the baffling array of odd varieties of the elementary particles.

Pais [6][27] has recently put forward a theory of heavy particles based on an extension of the manifold of transformations from four dimensions to six. While the idea of adding extra dimensions is by no means new, Pais' particular method is original and interesting. By means of it he is able to predict the existence of particles which have properties similar to, and which he identifies with, nucleons, V particles and π and τ mesons. Furthermore he is able to derive the charge independence of nuclear forces and the law of conservation of heavy particles from fundamental invariance conditions.

Although Pais specifically disclaims any metric properties for his extra two dimensions, there is no reason why his theory cannot be made completely geometrical. I should therefore like to suggest the following model for Nature: The universe is six-dimensional, with five space-like dimensions and one time-like dimension. Of the five space dimensions, two are closed on themselves with the topology of a spherical surface. The other three are the familiar dimensions of space. The reason we are not immediately aware of the two closed dimensions is that the spherical radius involved is extremely small. The metric of the six-dimensional manifold varies according to a set of equations derived from a *variational principle* based on the total curvature tensor. The metric tensor describes all *boson* fields, including the gravitational, electromagnetic, π and τ meson fields. The *fermion* fields, including electrons, neutrinos, muons, nucleons, and V particles, are described by a superimposed spinor field [7] together with a corresponding *lagrangian* function.

Preliminary investigation indicates that this model will possess all the features of Pais' model, including mass spectrum and stability properties. In addition it will yield necessary cross-couplings that Pais was unable to account for.

The final point of this note is now evident. If the gravitational field is welded into a single entity along with electromagnetic and meson fields, circumstances can arise (at least in the subnuclear domain) in which one field cannot be distinguished from another, and a broadening of the term

[27] Note added in proof by DeWitt: "In my reference to the work of Pais, I state that my proposal of a six-dimensional space-time is not only equivalent to, but is also a generalization of his ideas. This is not true, as I have discovered by further investigations carried out since I submitted this essay. My proposal is similar to Pais' at a number of points, but proves to be inadequate to describe the experimentally observed physical situation at a number of points."

"gravity" becomes inevitable. Under these circumstances one may will anticipate being able to "harness gravity." The vast riches of Nature in this domain are as yet virtually untouched.

If, however, one is ever to be able to do more than merely sit in contemplation of the delicate interplay of forces, both vast and small, between the elementary particles, one must understand in the clearest possible terms precisely what goes on behind the scenes. At the moment, our understanding of these matters is extremely poor. To the extent that this lack of understanding falls in the domain of gravitational theory (in the largest sense), the unrewarding nature of research in this field is to be blamed. External stimuli will be urgently needed in the near future to encourage young physicists to embark upon gravitational research in spite of the odds.

References

1. Pirani, F. A. E., Schild, A.: "On the Quantization of Einstein's Gravitational Field Equations," *Physical Review* **79**, 986 (1950)
2. Bergmann, Penfield, Schiller, and Zatzkis: "The Hamiltonian of the General Theory of Relativity with Electromagnetic Field," *Physical Review* **80**, 81 (1950)
3. DeWitt, B.S.: "Point Transformations in Quantum Mechanics," *Physical Review* **85**, 653 (1952)
4. See, for example, Bohm, D.: *Quantum Theory*, New York, Prentice-Hall, Inc. (1951), Chapter 22.
5. DeWitt, B.S.: "Pair Production by a Curved Metric," *Physical Review* **90**, 357 (1953) (abstract). For other cases, in which quantization of the gravitational field leads to no difficulties, references may be made to my Ph.D. thesis, Harvard (1950).
6. Pais, A.: *Proceedings of the Lorentz-Kamerlingh Onnes Conference*, Leyden, June 1953.
7. For the treatment of spinors in a unified field theory see Pauli, W.: "The formulation of the natural law with five homogeneous co-ordinations Part II The equation for the matter waves," *Annalen der Physik* **18**, 337 (1933). *See also* DeWitt, B.S., DeWitt, C.M.: "The Quantum Theory of Interacting Gravitational and Spinor Fields," *Physical Review* **87**, 116 (1952)."

DeWitt's interest in gravitation came to the attention of Agnew Bahnson, a Winston-Salem industrialist. Through the efforts of John Wheeler[28], Bahnson's interest and enthusiasm for antigravity devices were directed into a more fruitful pursuit: The creation of the Institute of Field Physics at the University of North Carolina in Chapel Hill N.C.

[28] An interesting letter from John A. Wheeler to Max Born, dated 31 March 1964, recalls the early negotiations that led to the creation of the Institute of Field Physics.

III.4 The Institute of Field Physics (1955–1964)

The story of the Institute began on May 30, 1955, with a letter from Agnew H. Bahnson, Jr. to Bryce DeWitt. It ended on June 3, 1964, when the plane Bahnson was piloting hit an electric wire and burst into flames. It is a tale that brings together unlikely partners:

- Agnew Bahnson, an industrialist from Winston-Salem, and Bryce DeWitt, a physicist from the Radiation Laboratory of the University of California (Berkeley and Livermore). Agnew, larger than life, with undaunted enthusiasm, Bryce, not restricted by fashion or traditions, but actually aware of obstacles.
- A private corporation, the Institute of Field Physics (Incorporated in the State of North Carolina on the 7th day of September 1955) with its own bylaws, and the University of North Carolina with its state-mandated structure.

For several years prior to 1955, Agnew Bahnson had been corresponding with George Rideout, President of the Gravity Research Foundation, who showed him Bryce's Gravity Research Foundation prize winning essay (reproduced in the previous section) and suggested that he contact Bryce.

In his letter to Bryce, Bahnson offered "to raise enough funds to give reasonable support to a scientist" who would "devote his attention primarily" to "basic research" in gravitation. A number of options were discussed for implementing Bahnson's offer and an Institute of Field Physics was established at the University of North Carolina at Chapel Hill.

The DeWitts (Bryce, Cécile, Nicolette, and by then a second daughter Jan) arrived in Chapel Hill NC in January 1956. Jan had been born in Dinubra CA, Bryce's family home town, on January 1st, 1955; she was the first baby of the year in Tulare County and her birth was announced on the local radio as "eight pounds of loveliness straight from heaven". Lovely she remains.

No short statement can summarize the nine years of intense activity of the Institute and the continuing support and interest of its founder. In the following section we include a few original documents, in Sect. III.4.2 we highlight a few achievements from those times.

III.4.1 Original Documents

- Agnew Bahnson's letter to Bryce DeWitt, dated May 30, 1955. Parts of the text have been underlined by Bryce DeWitt.

- The Memorial to Agnew H. Bahnson, Jr.
- A statement concerning the proposed Institute of Field Physics by Freeman J. Dyson dated October 1955.
- An undated handwritten document "The Challenge" by Bryce DeWitt (probably 1955).

The DeWitt files of the Institute of Field Physics have been deposited at the Center for American History of The University of Texas at Austin. Two items are selected from the DeWitt files as samples of issues of the times.

- An undated (probably 1959) memo by Cécile DeWitt "The combined leadership of private philanthropy and universities in strengthening American Science" (Excerpts).
- A memo dated February 21, 1964, from Cécile DeWitt to E.D. Palmatier, then Chairman of the Department of Physics, concerning the hiring of communists.

THE BAHNSON COMPANY

Complete Industrial Air Conditioning

1001 SO. MARSHALL ST.
WINSTON-SALEM, N. C.

TELEPHONE 4-1581
CABLE ADDRESS
"BAHNSON WINSTON SALEM"

May 30, 1955

Mr. Bryce S. DeWitt
University of California
Theoretical Division
Radiation Laboratory
Berkeley 4, California

My dear Mr. DeWitt:

For several years I have had correspondence with Mr. George Rideout, President of the Gravity Research Foundation in New Boston, New Hampshire. Recently, I wrote him about the fact that the Burlington Mills Company in Greensboro, North Carolina, with whom we have done a considerable amount of business over a period of thirty years, has given a two hundred thousand dollar grant to the State College of the University of North Carolina, at Raleigh, North Carolina, for the building of a nuclear reactor and other laboratory facilities. That laboratory was dedicated about a week ago. I had hoped to attend the dedication but was unable to do so. I did talk to Mr. Spencer Love, Chairman of the Board of Burlington Mills, who is a good friend of mine, and mentioned to him the thing that has been of interest to me for over twenty years. He seemed quite willing to investigate the possibility further in connection with the work in this nuclear laboratory at State College.

You may have recently heard of the division of Glenn L. Martin Aircraft company and I believe of the Convair Division of General Dynamics, that has set up research in anti-gravity as a new method of supporting heavier than air machines above the surface of the earth. You may have also read the article by Mr. William Lear of the Lear Radio Company in the last October issue of FLYING MAGAZINE predicting that fifty years from now the air plane would be a horse and buggy and that anti-gravitational reaction would support aircraft at any desired height above the earth. This may sound a little like the flying saucer deal but I believe it has a very practical opportunity of being worked out during our normal lifetime. Twenty years ago such ideas were not received with much hope of practical consummation. I recall discussing such things with Dr. David Griggs, who is now at the University of Southern California in the Geological Department. He lives in Brentwood which is on the outskirts of Los Angeles. I doubt if you have ever crossed his path but I am sending him a copy of this letter in the hope that you may have had some contact or may have such contact if you are ever in his vicinity or he is in yours.

I agree with your letter that the field of gravitation is quite unexplored. It seems to be one of the most important pioneer frontiers in science today. The

HUMIDIFYING · HEATING · VENTILATING · COOLING · AIR FILTERING · DEHUMIDIFYING

Mr. Bryce S. DeWitt
University of California

May 30, 1955

practical application of an anti-gravity aircraft, if it could be developed, would certainly change our whole concept of transportation, even more radically than the development of the automobile or airplane itself, in my opinion. It would probably also have broad repercussions in international relations and the entire concept of both trade and political associations between men all over the earth. One fearful note is seen in the accelerated development of weapons both in the nations of the free world and in the Communist dominated areas in that we are undoubtedly not alone in dreaming of such a mechanism and I believe it is a foregone conclusion that the Communist scientists are working along these lines today.

I note your proper feeling that a great deal of theoretical background must be given to the study of gravity before anything practical can be developed. I once had the pleasure of exchanging letters with Mr. Albert Einstein, specifically calling to his attention my hope of the eventual development of some anti-gravity aircraft and he very courteously wrote back the same point that as far as he was concerned, he did not feel confident to step from the theoretical into the practical. You may be of the same temperament or focus of interest and what I am now exploring may not be of interest to you. At the same time, at the suggestion of Mr. Rideout, I did want to lay a few hopes at your threshold for consideration.

What I would like to do, and I think the mechanics of doing it can be made practical in the near future, is to get a qualified scientist who would join the staff of the Neuclear Physics Laboratory at Raleigh, North Carolina, and devote his attention primarily to this phase of gravitational study. I have discussed this matter with Dr. Clifford Beck who is head of the Physics Department at State College and he seems to feel that it is quite workable. I believe I can raise enough funds to give reasonable support to such a scientist with a guaranteed position for at least five years. Dr. Beck made this suggestion because he said we would certainly not stumble on anything dramatic in this field of endeavor in any short period of time. I am quite sure that he agrees with your letter, that basic research must be done in the problem before we can turn our specific attention to the anti-gravitational aircraft project. They have 45 advance students and a rather complete staff available to co-operate with this work and they have a well-equipped laboratory.

I apologize for approaching you, Mr. DeWitt, with no more introduction or forewarning. I will appreciate your comments on this plan. Do you think that you would be interested personally in pursuing the matter further? If not, do you have anyone in mind that might fill such a position?

Mr. Bryce S. DeWitt
University of California

May 30, 1955

With best regards.

Sincerely yours,

Agnew H. Bahnson, Jr.

AHBjr/s

cc-Mr. George Rideout
-Mr. David Griggs
-Dr. Clifford Beck

Memorial to Agnew H. Bahnson Jr.

The Agnew H. Bahnson Jr. Professorship of Physics was announced on March 30th, 1965 and offered to Bryce S. DeWitt who held the Professorship until 1971.

The story of the Institute of Field Physics was marked by Bahnson's characteristic openness, directness and willingness to be corrected without losing momentum. The following excerpt from his writings may, however, give some insight into the spiritual quality of his drive.

"No one man lives long on this earth. It will take concerted efforts of many men to pry forth one of the deepest and most obstinate, but one of the most important and potentially useful secrets of nature...

In this quest we are reaching for the stars, – and beyond. Yet ... 'Lest a man's reach should exceed his grasp, then what's a heaven for?' ... A man's short life upon this earth is but the twinkling of an eye in cosmic time. If we look down the vistas of the past and glance ahead at the unknown eternity of the future, there is a frightful feeling of impotence in the contribution we can make to the world. ... Only spiritual values, aesthetic expressions, and the contribution to basic knowledge have an aspect of permanence in the civilization of man.

... It is the hope of this Institute that it may bring to many people the satisfaction of understanding better certain basic secrets of nature and the opportunity of ... penetrating one of the last and greatest strongholds of Nature which remains unassaulted with success."

- Agnew H. Bahnson, Jr.

THE INSTITUTE FOR ADVANCED STUDY
PRINCETON, NEW JERSEY

SCHOOL OF MATHEMATICS October 22, 1955

Statement concerning the proposed Institute of Field Physics.

I give my whole-hearted approval to the initiative of Mr. Bahnson in establishing an Institute of Field Physics at Chapel Hill, N. C. The main purpose of the Institute is to stimulate study and research into the gravitational field and its connections with other physical phenomena. This purpose is to my mind reasonable, timely, and forward-looking; if sufficiently sustained it has a chance of leading to major increases in our general scientific understanding, in quite unforseeable directions.
At the same time I wish to state certain conditions which I consider essential to a sound development of the Institute.

(1) The organisers must understand that a direct attack on the basic problems of gravitation is about the most formidably difficult task in the whole of science. It is unreasonable to expect that an attack of this kind should result in success within any fixed number of years. It is also unreasonable to expect that the members of the Institute should even attempt such an attack as a regular part of their duties.

(2) The actual program of the Institute will necessarily be much more modest. To begin with, the main effort should be to teach students and other scientists what is already known about gravitation, and to stimulate a wider interest in the subject.

(3) The Institute will succeed only if it is able to attract and hold first-rate men to its senior staff, and good students to its junior staff.

(4) To hold a first-rate man, it is necessary to offer a position with the status and advantages of a good University Professorship, including permanent security of tenure. This will require a very solid financial backing. For a permanent establishment of two or three professors, with ten student members, an endowment of one or two million dollars should be envisaged.

(5) To avoid becoming isolated and sterile, the Institute must operate as much as possible within the framework of normal university life. It is desirable that both staff and students of the Institute should have official status in the university to which the Institute is attached. For attracting good students, the most important requirement is that the University have a first-rate and many-sided activity in addition to the activity of the Institute.

If these five conditions can be met, I believe the Institute has a good chance of becoming a center of scientific progress, in which case it will ultimately repay a hundred fold the money invested in it. If these conditions are not met, the enterprise may easily become a dismal flop. It is a gamble which I consider well worth taking.

Freeman J. Dyson

Freeman J. Dyson
Professor of Physics

"The Challenge

Which Challenge?

Today the whole world is aware of the astounding growth and development of modern science. The world does not know, however, much about this growth – whether, for example, its present momentum will continue, which direction it will take next, or, perhaps most important, whether it is as broad in spirit, healthy, and well balanced as it might be. The very rapidity of scientific development can make us lose sight of the fact that certain problems of a very fundamental nature have yet to be tackled. To take an example from the domain of physics, one may call attention to the phenomenon of gravitation. In contrast to the study of electromagnetism, the consequences of which have been enormously fruitful, studies of gravitational phenomena have been sporadic and with few remarkable results. For many reasons the laboratories and research centers of the world are presently preoccupied with other matters, often the same matters, duplicating the work of each other, with identical equipment, while the phenomenon of gravitation is almost universally ignored.

Except for Newton and Einstein few physicists have pondered on this problem long or hard enough to make any progress. There is no doubt that the problem is especially difficult, and outwardly unrewarding. There are indications, however, that the time is again ripe for a new attack to be launched against the mysteries of gravitation. New mathematical tools developed in recent years, new and puzzling discoveries about elementary particles, the growing suspicion that one ought to begin again turning over some old conceptual stones, the knowledge that the consequences of present theories of gravitation are far from having been explored to their logical conclusions; these are but some of the signs."

On "The combined leadership of private philanthropy and universities in strengthening American Science".

Cécile DeWitt analyzes the leadership of private philanthropy in the light of the December 1958 report of the President's Science Advisory Committee under the chairmanship of James R. Killian, Jr. "Vital" and "unique" are the qualifications of private philanthropy which comes as a leitmotiv in the government report.

Then the author analyzes the leadership of Universities: "ultimately the growth of science rests on the relationship between research and education and on the relationship between the Sciences and the Humanities". The importance of this relationship is "not obvious but (is) as vital as underground waters ... Universities enjoy a vantage point from which to examine the many aspects of the problems raised by research, education, the sciences, the humanities and their relationships. They have both unexcelled opportunities and qualified personnel to work out forward looking solutions to these problems."

C. DeWitt presents the Institute of Field Physics as an example of the benefits from a combined leadership of universities and private philanthropy. "In January 1956 the Institute of Field Physics was founded through contributions of private individuals and companies to promote research in gravitation, an important but neglected area in physics in which progress could be expected only if there was an opportunity to make a prolonged and concentrated effort on its study. The Institute, an integral part of the Department of Physics, consists of a small staff working actively on gravitation and related problems. The progress made in the understanding of this universal phenomenon may be assessed by comparing the proceedings on an international conference sponsored by the Institute of Field Physics in January 1957 at Chapel Hill and the proceedings of a follow-up conference held in June 1959 at Royaumont (France). The proceedings of Chapel Hill present essentially a picture of the work to be done. The proceedings of Royaumont show the first attacks on the problems, new tools and techniques which can make possible the solutions of the problems discussed two years earlier, and a partial understanding of the basic notions underlying Einstein's theory."

"The history of the Institute of Field Physics displays the features of a successful collaboration between a university and a private initiative: Mr. Agnew H. Bahnson, Jr., of Winston-Salem, who for a number of years had been puzzled by the phenomenon of gravity, made contacts with the persons likely to give him fruitful advice on the means to foster research in gravitation. In the course of his inquiries, it became clear to him that a university is the best place for such a project to grow and that providing suitable working conditions to persons who have the interest and ability to study the problem is the best step towards its solution. He proceeded to raise the money for a initial three-year budget. This budget has with judicious use of national and Federal resources been expanded and extended for another two-year period. If either state or private resources or a combination of both can guarantee the permanent establishment of the Institute, national and Federal resources will continue at a rate sufficient to insure a rich program of activities and to make possible the full development of this endeavor which is already bearing its first fruits".

The witch-hunt against communists

The witch-hunt against communists and communist sympathizers was in full swing during this period. The following memo, dated February 21, 1964, from Cécile DeWitt to E.D. Palmatier, then chairman of the Department of Physics of the University of North Carolina is but one item that illustrates this topic; it deals with the exact formulation of the rule stating that the University does not knowingly hire a Communist.

"In view of the fact that many outstanding theoretical physicists are from countries from behind the Iron Curtain, and in view of the fact that the Institute of Field Physics would benefit greatly from discussions with those physicists, it would be very helpful to have an exact statement of the rule mentioned above and an interpretation of its application with regard to the invitation for foreigners for short visits. I am listing below some questions which have come to our minds in this regard.

1. Exact statement of the law. Affiliation to the Communist Party. Does that mean present affiliation or does it cover past affiliation?
2. "Knowingly". – Which inquiries do we have to make? Obviously, we cannot ask the question directly to someone who is behind the Iron Curtain. When a foreigner applies for a visa, he is checked in a way that American citizens are never checked. Can the issuance of a visa be considered a satisfactory proof of the acceptability of the person?
3. Hiring. What constitutes hiring? Is a visitor who has no faculty privileges considered as "hired"? Does a short visit constitute hiring? If not, what length of time is considered as short?
4. In the past on the PD-7 form there was a statement to the effect that the person was not a Communist. Foreigners did not sign that statement and instead of the signature marked "does not apply". The statement was removed several years ago. It does not seem that the hiring of foreigners would be more restricted now than it used to be in the days when the statement was in effect.

Strong efforts are made to have the speaker ban law repealed. This would then imply that it is more restrictive than the law stated above. In which sense is the above law less restrictive than the speaker ban law?"

A clipping of the Daily Tar Heel, the university's daily newspaper, entitled "Friday Says No Red Profs On Campus" and attached to the memo reads: "Consolidated University President William C. Friday has challenged critics of UNC to prove that there are any Communist party members on the faculty of its campuses."

The following story may serve as an anecdote in this context: Leon Rosenfeld who had been invited to the conference on "The Role of Gravitation in Physics," organized by the Institute to be held in Chapel Hill from

January 18 to 23, 1957, had been denied a visa to enter the US by the American consul in Manchester on the grounds that he was a Communist sympathizer. Upon discovering that the only authority who could reverse a consul's decision was the US Attorney General, Cécile called him directly, simply getting his phone number from Directory Assistance. With a self-assured tone of voice she was quickly transferred by his secretary to the Attorney General, who immediately granted her request. Different times, different customs.

4.2 A Stone Thrown Into a Pool and Its Ripples

One of Bryce DeWitt's favorite quotes from Rudyard Kipling's novel *Kim* speaks of the unforeseeable ripples an action engenders: "Thou has loosed an Act upon the world, and as a stone thrown into a pool so spread the consequences thou canst not tell how far".[29]

The following lists of The Institute of Field Physics personnel, Ph.D. graduates, visitors, and Texaco fellows illustrate some of the ripples triggered by Bahnson's actions.

Permanent:
Bryce S. DeWitt, Director of Research
Cécile M. DeWitt, Visiting Research Professor
Hendrick Van Dam, Assistant Professor (appointed January 1962)

Temporary appointments – Research Associate:
B. E. Laurent, Institute of Theoretical Physics, Stockholm, 1957–1958
F. A. E. Pirani, Kings College, London, 1958–1959
T. Imamura, Osaka University, 1958–1960
I. Robinson, University of Aberstwyth, 1959–1960
L. E. Halpern, Institute of Theoretical Physics, Vienna, 1960–1961
G. W. Erickson, University of Minnesota, 1960–1962
F. R. Tangherlini, University of Naples, 1960–1961
R. Utiyama, University of Osaka, 1960–1961
H. Van Dam, Foundation for Fundamental Research on Matter, Utrecht, 1961–1962; 1963–1964 (half-time); 1964–1967
T. W. Noonan, Astrophysical Obs., Cambridge, MA., 1962–1963 (half-time)
J. M. Knight, Duke University, April, 1964 December, 1964
E. A. Remler, University of North Carolina, 1963–1964
F. Karolyhazy, R. Eötvös University, Budapest, Hungary, 1963–1965
Joanna Zund, Southwest Center for Advanced Studies, Dallas, Texas, 1964–1965
Joe Zund, University of Texas, Austin, 1964–1965
Robert W. Brehme, Wake-Forest College, Winston-Salem, NC., 1964–1965

[29] Rudyard Kipling, "*Kim*". MacMillan, London, (1950), pp 271–272.

G. Papini, University of Catania, Italy, 1964–1966
P. Higgs, The University of Edinburgh, Scotland, 1965–1966
H. Pagels, Stanford University, Stanford, California, 1965–1966
P. Droz-Vincent, Centre National de la Recherche Scientifique, Paris, 1965
M. Miketinac, Yugoslavia, 1965–1967
G. Braunss, University of Darmstadt
R. Utiyama, University of Osaka, 1966–1967
T. Imamura, Kwansei Gakuin University, 1966–1967
L. Parker, Harvard University, Cambridge, MA., 1966-1967 (Instructor)
M. Dillard, University of North Carolina, Chapel Hill, NC., 1968–1969
F. Zerilli, Princeton University, Princeton, New Jersey, 1969

Ph.D. graduates:
Robert W. Brehme, "A Charged Particle in a Static Gravitational Field", 1959
John J. Ging, "Gravitational Radiation Damping", 1960
Hsin Yang Yeh, "Quantum Limitations on the Measurability of the Gravitational Field", 1960
Edward A. Remler, "Cross Sections for Yang-Mills Quanta", 1963

Fig. 9 Bryce and Ivor Robinson (then from the University Aberystwyth) at the Institute of Field Physics (1959–1960). The long list of members of the Institute reflects the intense activity at the Institute during its existence from 1955 to 1964

Charles F. Cooke, “Gravitational Scattering Cross Sections”, 1964
Allen C. Dotson,“Quantum Theory of Interacting Gravitational and Yang-Mills Fields”, 1964
William C. Rodgers, “Computations of Orbits and the Light Cone in Schwarzschild Field”, 1964
John L. Safko, “Peratization Methods in the Quantum Theory of Gravity”, 1965
James H. Cooke. “Quantization of Relativistic Action-at-a Distance Theories”, 1966
Margaret A. Bleick Dillard (Dillard-Bleick)[30], “Tensor and Spinor Harmonics on the 5-Sphere”, 1967
Peter B. Eby, “Classical Relativistic Mechanics of Particles with Spin”, 1968
Edith Borie, “The S-Matrix in the Heisenberg Representation”, 1968
Milivoj J. Miketinac, “A Nine-Dimensional Unified Field Theory making Use of the 5-Field Theory”, 1969
Andrej Cadez, “Computer Calculations of Two Colliding Black Holes”, 1971
Roger Neill Graham, “The Everett Interpretation of Quantum Mechanics”, 1971
Thomas W. Hill, “Cosmological Computations Using Quantum Corrections to Einstein’s Field Equations”, 1971
Walter G. Wesley, “Quantum Falling Charges,” 1971

Visiting Speakers:
O. Klein, Institute of Theoretical Physics, Stockholm, April-May 1958
C. Møller, Nordita, Copenhagen, February 1958
B. Bertotti, Institute for Advanced Study, Princeton, April 1959
A. Komar, Syracuse University, May 1959
G. Rosen, Princeton University, April 1959
R. Karplus, University of California, March 1959
D. Sciama, Cambridge University, June 1961
F. Gürsey, Institute for Advanced Study, Princeton, February 1964
N. Rosen, Technion, Haifa, Israel, September–October 1962
J. Plebanski, University of Mexico, November 1963
L. Michel, Institut des Hautes Études Scientifiques, Bures-sur-Yvette, France, January 1965
R. H. Dicke, Princeton University, April 1966
L. Parker, Harvard University, May 1966

Texaco fellows:
William C. Rogers, 1961–1963
Charles F. Cooke, 1963–1964
Walter G. Wesley, 1964–1966
Peter B. Eby, 1966–1967

[30] as contributor to *Analysis, Manifolds and Physics: Part I, see* footnote 17 in Sect. I.1.

A well-known ripple: The Higgs boson

The list of publications of the Institute of Field Physics includes many papers whose ripples were felt far and wide. The paper by Peter W. Higgs entitled "Spontaneous Symmetry Breakdown Without Massless Bosons" (*Phys. Rev.* **145**, 1156–1163 (1966)) launched the concept known nowadays as the Higgs Boson.

The 1957 Chapel Hill Conference (*see* Sect. III.5) was followed by a General Relativity and Gravitation (GRG) conference in Royaumont in 1959 and one in Warsaw in 1962. These three conferences are known as GR1, GR2, and GR3, respectively, in the GRn series. Bryce DeWitt had met Peter Higgs at the GR2 and GR3 conferences[31] and invited him to spend one year at the Institute of Field Physics. Peter Higgs arrived in Chapel Hill in September 1965.

The Higgs mechanism The Higgs mechanism implies the existence of a particle that became known as the Higgs Boson. It makes Quantum Mechanical predictions in the Nambu program from symmetry breaking in particle physics. But it does more than that: it provides a useful bundle-reduction-example[32] in the mathematical theory of fibre bundles, the theory that nowadays underlies Quantum Field Theory.

The symmetry-breaking process alluded to above is the spontaneous symmetry breaking process that occurs in a system when the state of lowest energy (called "vacuum") has fewer symmetries than the lagrangian of the system.

The 2004 Wolf Prize Thanks to the Institute of Field Physics preprint distributions, Peter Higgs was invited to give seminars at the Institute for Advanced Study (March 15, 1966) and at Harvard (March 16, 1966) even before his 1966 paper that appeared on 27 May 1966 – his work was promptly discussed.

In due time, recognition has also been given to other physicists who had proposed a mass generating process identical to the Higgs mechanism. Physical Review Letters 50th anniversary celebration included several paper related to this process. And in 2004 the Wolf prize was attributed jointly to Robert Brout, Francois Englert, and Peter Higgs "for developing the theories

[31] A letter from Peter Higgs to Cécile DeWitt dated 27 August 2009 tells the story of his 1966 paper and his interactions with Herman Bondi, Felix Pirani, and Abdus Salam. This letter is stored in the archives of the Center for American History (*see* Sect. V.4).

[32] Y. Choquet-Bruhat and C. DeWitt-Morette, *Analysis, Manifolds, and Physics: Part II.* North Holland, Amsterdam (2000), pp 310–321.

which explain how fundamental particles can acquire mass in the Standard Model of Particle Physics".[33]

Regarding his time spent at Chapel Hill, Peter Higgs writes in his letter to the author[31]: "All of this, together with the presence of another particle theorist (Heinz Pagels), resulted in my not making any contribution to Quantum Gravity during my time at Chapel Hill. Bryce must have been very disappointed."

A landmark ripple: The Falling-Charge Problem

The role played by the Ph.D. thesis of the first graduate student at the Institute, Robert W. Brehme, can be found in DeWitt's contribution to the book dedicated to Schwinger's memory.[34] In the mid fifties DeWitt was trying to develop the canonical formalism[35] for quantizing the gravitational field. At Brehme's request he considered a problem that led him to the covariant formalism that became his formalism of choice.

"I owe a considerable debt to my first student, Robert Brehme, who, in the late 1950's, asked me whether he could work on the problem: Does the equivalence principle apply to charged matter? I did not at first regard this as a very interesting problem. I had immediately translated it in my mind to the question: Does a falling charge radiate? And I saw no reason why it should not. In my view the issues of the equivalence principle was a red herring. The principle was never meant to apply other than locally to physical phenomena, and a charged particle is hardly a local object in view of the extended Coulomb field that it carries with it.

I was at that time trying to develop a canonical formalism for the gravitational field with the aim of creating a quantum theory of gravity, and I hoped that Brehme would assist me in this work. In fact, the work bogged down in the usual difficulties familiar to anyone who has tried to construct, and make sense of, a canonical quantum theory of gravity. So, in desperation, I agreed to let Brehme investigate the falling-charge problem; but I insisted that he do it properly. He was to begin by studying Dirac's famous 1938 paper on

[33] Belle Dumé, Science Writer at *PhysicsWeb*: "Jan 20, 2004 Wolf Prize goes to Particle Theorists" ("http://physicsworld.com/cws/article/news/18884"). The official citation states that the awards are made "for pioneering work that has led to the insight of mass generation, whenever a local gauge symmetry is realized asymmetrically in the world of subatomic particles." P. W. Higgs "Spontaneous Symmetry Breakdown Without Massless Bosons" *Phys. Rev.* **145**, 1156–1163 (1966); R. Brout and F. Englert "Broken Symmetry and the Mass of Gauge Vector Bosons" *Phys. Rev. Lett.* **13**, 321–323 (1964).

[34] [BD 93b] This paper is an excellent introduction to DeWitt's works (pp. 34, 35, 42, 43, 46, 47, 49, 50).

[35] *see* [BD 2] and [BD 31] – his most cited paper, and the only reference given in Wikipedia for their entry on the "Wheeler–DeWitt equation."

the classical radiating electron, in which all calculations are performed in a manifestly Lorentz covariant manner. He was then to translate this paper into the language of curved spacetime, keeping all the derivations manifestly generally covariant. He was not to introduce a special coordinate system at any stage.

The first obstacle he encountered was the problem of wave propagation in curved spacetime. Nobody seemed to have looked at this problem, at least in the physics literature. At length I discovered its solution, or at least part of its solution, in Hadamard's book *Lectures on Cauchy's Problem in Linear Partial Differential Equations* (Yale, 1923). In this book Hadamard does not use covariant notation or terminology, but it is easy to recast his results into covariant form. ... My debt to Robert Brehme is not limited to the fact that his thesis work led to my learning ... lovely results. I have also learned other lovely results by reflecting on the general properties of Green's functions, not only for boson fields but for fermion fields as well. ... The reciprocity relations [of Green's functions] were of interest to me because I had encountered them at about the same time (1960) in an entirely different context. I was making a great effort to study the famous but difficult paper by Bohr and Rosenfeld[36] the measurability of the quantized electromagnetic field, with the aim of settling the analogous question whether there can be any measurement-theoretical meaning to quantizing the gravitational field. The chief issue concerns the effect of disturbances caused by the measurement process itself."

The reciprocity relations made it possible for DeWitt to construct more general Poisson brackets, first presented by Peierls[37]:

"The remarkable thing about Peierls' brackets is that they do not depend for their definition on the introduction of a canonical formalism. They are completely determined by the laws of propagation of Jacobi fields, and their definition emphasizes a global spacetime view of the dynamics. When I first realized that Bohr and Rosenfeld were dealing with Peierls brackets,

---

[36] N. Bohr and L. Rosenfeld, *Kgl. Danske Videnskab. Selskab, Mat.-fys. Med.* **12**(8) (1933). The Bohr and Rosenfeld paper is in German – it has been translated into English by DeWitt in 1960. A type-written copy of the translation is in the Library of the Institute for Advanced Study (Princeton, NJ). In a letter to Cécile DeWitt, dated 30 October 2004, F. J. Dyson wrote "Bryce did a beautiful job of translating Bohr's convoluted German sentences. He must have worked hard on it, as it goes on for sixty-three pages. ... I wonder whether this was ever published. If not, it should be. It is a classic in the history of physics, and ought to be accessible to people who do not read German. It would also be a good memorial for Bryce." A first unsatisfactory attempt at the translation had been made by Cécile DeWitt and was best discarded.

[37] R.E. Peierls "The commutation laws of relativistic field theory", *Proc. Roy. Soc. (London) A* **214**: 143–157 (1952).

I became quite excited. . . . The Peierls bracket is seen to have the same symmetry as the supercommutator[38] bracket in the quantum theory. Therefore the Peierls bracket is the appropriate concept for analyzing the quantum mechanical limitations on measurement accuracy. This analysis says that measurements can, in principle, always be made to an accuracy equal to *but no better than* that allowed by the *a priori* uncertainties implied by the quantum mechanical formalism. . . . Thus, quantizing the gravitational field is exactly as meaningful as quantizing the electromagnetic field. There is just one new limitation, which does not exist in the electromagnetic case: The sizes of the spacetime averaging domains must be large compared to the Planck length, 10^{-33} cm. Many arguments lead to the conclusion that standard notions of space and time, and even of probability itself, cease to have operational meaning below this scale. This is the domain in which string theory is supposed to have something new to say.

My excitement over the discovery of the role of the Peierls bracket in the Bohr–Rosenfeld analysis stemmed from the fact that Bohr and Rosenfeld, and also myself in the gravitational case, were able to confine our attention exclusively to observable[s]. . . . One beautiful application of the Peierls bracket that is of fundamental practical importance is its use in the derivation of the Schwinger variational principle that leads to a Feynman functional integral with many practical applications. Little did I know in 1959 that Robert Brehme's problem would lead to all this."

A ripple felt far and wide: The Wheeler–DeWitt equation

Section III.2.2 includes an extensive abstract from DeWitt's own article on the Wheeler–DeWitt equation written in 1996. A brief mention of the Wheeler–DeWitt equation appears in Weinberg's biographical memoir that is reproduced *in extenso* in Sect. V.1.

Guggenheim Fellowship

The Institute of Field Physics ceased to exist upon the death of Agnew Bahnson in 1964. Bryce DeWitt remained at the University of North Carolina in Chapel Hill until 1971 as the Agnew H Bahnson Jr. Professor of Physics. A summary of his research while in Chapel Hill can be found in his application for a Guggenheim Fellowship (awarded 1975–76). It is reproduced below:

"1. In 1959 I discovered generally covariant Green's functions. By that I mean I discovered them for myself. Covariant Green's functions were

[38] *see* Sect. III.9 and [BD 91].

already implicit in the work of Hadamard decades before. But they had never previously been used consciously as such. I explored their properties in some depth and applied them to a variety of problems both classical and quantum mechanical. They completely transformed my view of and approach to the quantum theory of gravity.

2. With the aid of covariant Green's functions I made, in 1962, the first complete analysis of the quantum theory of measurement of the gravitational field, along the lines of the Bohr-Rosenfeld analysis of the same problem for the electromagnetic field.
3. In 1963 I attempted a grand synthesis of the Green's function approach to the quantum theory of non-Abelian gauge fields (of which the gravitational field is one). This resulted in a series of lectures at the Ecole d'Eté de Physique Théorique, Les Houches and in a book, *Dynamic Theory of groups and Fields*, based on those lectures. This book was little read at the time, but in recent years it has been discovered by particle physicists, and both its notation and its general point of view have been increasingly adopted.
4. In 1964 I succeeded for the first time in extending the Feynman rules for non-Abelian gauge theories beyond single closed loops. In 1966 this extension was generalized to all orders and expressed as a functional integral that includes as a limiting case the rules found independently by Faddeev and Popov in 1967. A well-defined quantum theory of gravity was born at last.
5. During 1966 to 1970, at the gentle prodding of John Wheeler, I had a fresh look at the canonical approach to the quantum theory of gravity, which I had abandoned years before.[39] This resulted in a new geometrical analysis of Superspace and in my proposing a functional differential equation as the foundation of the theory. The ideas developed in this work were applied by myself, and in modified form by others, to cosmological models with finite numbers of degrees of freedom.
6. In 1970 and 1971 I made, with my student Neill Graham, an exhaustive study of the literature on the interpretation of quantum mechanics, and I wrote several articles popularizing the Many-Universes Interpretation first proposed in 1957 by a Princeton graduate student, Hugh Everett. This interpretation seems to be the only one fully capable of embracing a quantum theory of gravity.

In 1971, my wife was offered a tenure position (the first time she had been able to secure one in many years) in the Astronomy Department of The

[39] *see* Sect III.2.2 on Covariant vs. Canonical quantization.

University of Texas. I therefore left the University of North Carolina and in the spring of 1972 took up my present position in the Relativity Center of The University of Texas."

I.5 The 1957 Chapel Hill Conference

In 1957 the Institute of Field Physics organized a conference on "The Role of Gravitation in Physics." It has become a landmark in the pursuit of quantum gravity, in part because the history of Hugh Everett III and the many worlds interpretation of quantum physics began there; and in part because it launched a series of international conferences on general relativity and gravitation. Prior to the Chapel Hill conference there had been a Jubilee of Relativity Theory held in Bern (Switzerland) in 1955. In the "GRn" series of conferences, it became known as GR0, with the 1957 Chapel Hill meeting being GR1; GR18 was held in Sydney in 2007.[40]

There are two reports of the Chapel Hill conference. An Air Force report was written by Cécile dated March 18, 1957[41]; its cover is reproduced here. This report endeavors to record the discussions, confusions, opinions, and hopes expressed during the conference as well as the presentations of the participants.

What became known as "Numerical Relativity" (*see* Sect. II.1) is introduced by DeWitt in Session 2 (pp 18–27 of the report) in the discussions following the presentations by C. W. Misner and by Yvonne Fourès (Yvonne Bruhat). Many other comments recorded in this report have made the 1957 Chapel Hill Conference a unique reference for quantum gravity.

The second report by Bryce can be found in the July 1957 issue of *Reviews of Modern Physics* **99**: 351–546. It assembles papers on gravitation that were ready for publication whether or not they had been presented at the conference. Bryce DeWitt's Introductory Note and the table of contents are reproduced here. The contents include Hugh Everett III's Ph.D. thesis, submitted to Princeton University on March 1, 1957, "Relative State Formulation of Quantum Mechanics"; it appears on pp 454–562. In Sect. III.6 the reader will find John Wheeler's comments on the 1957 Chapel Hill conference. Dean Rickles has used the Air Force report to write a book *The Role of Gravitation in Physics: Report from the 1957 Chapel Hill Conference*. It will appear in the Edition Open Access of the Max Planck Institute for the History of Science (Eds Cécile M. DeWitt, Dean Rickles, 2011).

---

[40] For the complete listing of the GR*n* conferences, look up http://grg.maths.qmul.ac.uk/grgsoc/conferences.html.

[41] Her third daughter, Christiane, was born on March 19 – the day after the report was mailed.

WADC TECHNICAL REPORT 57-216
ASTIA DOCUMENT No. AD 118180

CONFERENCE
ON
THE ROLE OF GRAVITATION IN PHYSICS

AT

THE UNIVERSITY OF NORTH CAROLINA, CHAPEL HILL

JANUARY 18-23, 1957

MARCH 1957

WRIGHT AIR DEVELOPMENT CENTER

Papers from the Conference on the Role of Gravitation in Physics Held at the University of North Carolina, Chapel Hill, North Carolina, January 18–23, 1957

Introductory Note

THE following papers were prepared in connection with the Conference on the Role of Gravitation in Physics which was held in Chapel Hill, January 18–23, 1957. The conference was initiated by the North Carolina Project of the Institute of Field Physics, Inc., established in 1956 in the Department of Physics of the University of North Carolina, Chapel Hill. The sponsors of the conference were the International Union of Pure and Applied Science (with a financial contribution from UNESCO), the National Science Foundation, the Wright Air Development Center, and the Office of Ordnance Research. The conference was organized—with the collaboration of the Institute of Natural Science of the University of North Carolina—by the following steering committee: F. J. Belinfante, Purdue University; P. G. Bergmann, Syracuse University; B. S. DeWitt, University of North Carolina; Cecile M. DeWitt, University of North Carolina; F. J. Dyson, Institute for Advanced Study; and J. A. Wheeler, Princeton University. It is a pleasure to mention the interest shown in the conference by the officers of the University, by the Governor of the State of North Carolina, who welcomed the physicists to Chapel Hill at a luncheon in the Morehead Planetarium, and by all who have made the Chapel Hill conference possible and have helped in its organization.

The conference was planned as a working session to discuss problems in the theory of gravitation which recently have received attention. A report of the proceedings of the conference can be obtained from Wright Air Development Center, Wright-Patterson Air Force Base, Ohio. The WADC report endeavors to record the discussions, confusions, opinions, and hopes expressed during the conference. A separate need was felt to assemble, in an issue of a scientific journal, the papers on gravitation which were ready for publication—whether or not they had been presented at the conference. Participants in the conference are grateful to the editors of *Reviews of Modern Physics* for making available the space necessary for these papers.

This collection of papers cannot strictly be considered as a review of present knowledge about gravitation. A noticeable increase of interest and activity in the theory of gravitation and related matters has taken place only recently, following a period of relative quiet. The problems attacked have not yet settled down so that they cannot be viewed with any kind of real perspective. At present the most that can be hoped for is that these papers will give a fairly reasonable picture of current activity in the field of relativity physics and stimulate greater awareness of a whole series of fundamental questions which have yet to be answered.

BRYCE S. DEWITT

Papers from the Conference on the Role of Gravitation in Physics Held at the University of North Carolina, Chapel Hill, North Carolina, January 18–23, 1957

1.6 Everett's Theory and the "Many Worlds" Interpretation[42]

As a tool, quantum mechanics is extremely powerful. Its conceptual foundation, on the other hand, is still subject to intense debates. Hugh Everett III began his first published paper[43] with these words:

"The task of quantizing general relativity raises serious questions about the meaning of the present formulation and interpretation of quantum mechanics when applied to so fundamental a structure as space-time geometry itself. This paper seeks to clarify the foundations of quantum mechanics. It presents a reformulation of quantum theory in a form believed suitable for application to general relativity.

The aim is not to deny or contradict the conventional formulation of quantum theory, which has demonstrated its usefulness in an overwhelming variety of problems, but rather to supply a new, more general and complete formulation, from which the conventional interpretation can be deduced."

Bryce DeWitt played a major role in bringing the many worlds interpretation of quantum mechanics to the attention of the physics community.[44] Two documents, one from Bryce DeWitt and one from John Wheeler, give a good picture of the debates generated by Everett's theory, and are included below. They also make it abundantly clear that one cannot summarize Everett's ideas in a few words.[45] Nevertheless, as a stepping stone to read the DeWitt and Wheeler documents, here is a rough introduction to Everett's theory, in DeWitt's words[46]:

"Everett's aim was to cut through the fuzzy thinking displayed by many authors, some of them quite prominent, who in previous years had written incredibly dull papers on how they understood quantum mechanics. Everett's idea was simply to assume that quantum mechanics provides a description of reality in exactly the same sense as classical mechanics was once thought to do."

---

[42] [BD 104, 97, 35, 40, 41].
[43] "Relative State Formulation of Quantum Mechanics", *Review of Modern Physics* **29**, 454–462 (1957).
[44] In 2002, Bryce DeWitt gave a talk on the topic at the Sharif Institute of Technology in Teheran, Iran. The announcement is reproduced in this section and the translation of the text reads: Sharif Institute of Technology – Lectures of the Department of Physics – Sunday 5th of Aban 1381 (27 Oct. 2002) – 16:30 afternoon – Department of Physics Amphitheater.
[45] The book by Thibault Damour "*Once upon Einstein*", AK Peters, Wellesley, MA (2006) may help.
[46] [BD 104] p 167.

سخنرانیهای دانشکده فیزیک

یک شنبه ۵ آبان ماه ۱۳۸۱
ساعت ۱۶:۳۰ بعدازظهر
دانشکده فیزیک، آمفی تئاتر

The Many-Worlds Interpretation of Quantum Mechanics

Professor B. Dewitt

University of Texas

U.S.A

The foundation of both classical and quantum physics rests on the analysis of measurements of physical systems. "In its simplest form a measurement involves just two dynamical entities: a system and an apparatus. It is the role of the apparatus to record the value of some system observables." A measurement is a disturbance in the combined system-apparatus history.

The study of measurements and disturbances has a long history in the Quantum Theory. Two landmarks in DeWitt's investigation of field quantization are:

- His translation of the Bohr and Rosenfeld 1933 paper[47] on the question of the measurability of the electromagnetic field strengths.
- His study of the Peierls' bracket [BD 103, 100, 105] for infinitesimal disturbances. It was originally introduced in 1952 by Peierls to give an alternative, non-canonical, covariant, global definition of the Poisson bracket. The Peierls' bracket is the cornerstone of DeWitt's field quantization.

In quantum physics one can only talk of the possible outcomes of a measurement. But then what can one say after a measurement selects one of these outcomes? What happens to every other possible outcome? According to the "Copenhagen Interpretation", the other potential outcomes vanish by necessity once a measurement has been made. For Everett the other potential outcomes are simultaneously realized within the universal wave function that describes quantum reality. An observer can only experience one reality at a time and all other possible realities exist in parallel universes. "This is a shocking idea [...] and few physicists in 1957 were prepared to accept it. [...] yet it can be shown to work."[48]

There are still loose ends to be tied up to complete Everett's vision. Here are three identified by DeWitt:

- An analysis of the Einstein-Podolsky-Rosen experiment as seen from Everett's viewpoint.
- A textbook setting forth a conceptual framework for entanglement and quantum computers.
- The Wheeler-DeWitt equation as seen from Everett's solidly grounded ideas.

[47] N. Bohr and L. Rosenfeld, *Kgl. Danske Videnskab. Selskab, Mat.-fys. Med.* **12**(8) (1933), *see* footnote 36 in Sect. III.4.
[48] [BD 104] p. 197.

III.6.1 Everett's Theory Viewed by Bryce DeWitt

On the occasion of writing a referee's report on an article submitted for publication, DeWitt set the historical record straight regarding Everett's work[49]:

"In January 1957 my wife and I hosted GRG1, the first international conference on General Relativity and Gravitation[50], at the University of North Carolina. Shortly thereafter the Reviews of Modern Physics agreed to publish a set of papers submitted by conference participants dealing with the topics they had discussed at the conference. The papers were to be submitted to me, as acting editor for the set.

One of the papers that I received was Everett's "Relative State" paper. Although Everett had not been a conference participant and I had never met him, his paper was accompanied by (1) a strong letter from John Wheeler urging acceptance and (2) a paper by Wheeler assessing Everett's ideas. Since Wheeler had been a very active conference participant and since Everett's paper seemed to be relevant to the themes of the conference, I agreed to include it. This, of course, meant that I had to read it. In retrospect it seems to me likely that at that time, and for a number of years afterward, Wheeler and I were the only people, besides Everett himself, who know what was in the paper. I read it very carefully and have a vivid memory of my reaction. First, I was tickled to death that someone at long last, after so many years and so many tiresome articles, had something new and refreshing to say about the interpretation of quantum mechanics. Second, I was deeply shocked.

I was so shocked that I sat down and wrote what turned out to be an eleven pages[51] letter to Everett, alternately praising and damning him. My damning largely consisted of quoting from Heisenberg regarding the "transition from the possible to the actual" and insisting upon the fact that "I do not feel myself split." The response I got from Everett was a Note to be added in proof to his article.

Before discussing this Note let me establish some conventions that will make referencing easy. Since all the papers to which I shall need to refer have been collected in "The Many-Worlds Interpretation of Quantum Mechanics," eds. DeWitt and Graham (Princeton, 1973), I shall use the pagination of that volume together with the letters MW. The coordinates of the Note in question are then (MW, bottom pp. 146, 147).

---

[49] We refer the reader interested in Everett and his theory also to the following publication: Peter Byrne, *The Many Worlds of Hugh Everett III*. Oxford University Press, Oxford (2010).

[50] Nowadays numbered GR1; *see* Sect. III.4.

[51] Eight typewritten pages. This letter is featured in the *APS News* **18**, 2 (2009) in the Department "This Month in Physics History" under the title "May 31, 1957: DeWitt's Letter on Everett's 'Many Worlds' Theory." W.W. Norton & Company New York, 1998.

The Note that Everett submitted was accepted by me and appeared in the published version. In it the word "splitting" appears. The word first appeared in my eleven page letter, but Everett accepted it without apparent qualms. He compared those who objected to his interpretation on the grounds that they don't feel themselves split, to the anti-Copernicans in the time of Galileo who did not feel the earth move. He also said " ... *all* elements of a superposition (all 'branches') are 'actual,' none any more 'real' than the rest." He might equally have said " ... any less 'real' than the rest," and he *did* say this elsewhere (MW top p. 107).

His reference to the anti-Copernicans left me with nothing to say but "Touché!" His reply to my letter was succinct and to the point. I had no further ammunition to throw at him. There the matter might have rested had I not received a visit from Max Jammer a few years later. Jammer was preparing another volume on the foundations of quantum mechanics, attempting to bring the historical record up as far as the 1960s. I was surprised to discover that Jammer had never learned of Everett. As I was to learn later, Everett himself couldn't have cared less. As a graduate student he had puzzled over the foundations of quantum mechanics, had found an interpretation that satisfied him, and had then dropped the subject, going to work for the Department of Defense upon receipt of his degree (MW, bottom p. 141) and later forming his own consulting firm. I shall comment on some of his subsequent interests presently.

The meeting with Jammer started me thinking: This young man (Everett) is getting a raw deal; something should be done about it. It happened that I had a graduate student (Graham) at the time whom I had (reluctantly) permitted to start a thesis on the foundations of quantum mechanics. I insisted that he have a close look at Everett. My initial shock at Everett's ideas had long since worn off, and whenever I thought about foundation questions (which was rarely) I found myself turning to Everett as providing the most sensible framework. As a result of discussions with Graham I resolved to start a publicity campaign. My initial act was to lecture on the Everett interpretation at the Battelle Recontres in Seattle in 1967. Then I wrote the *Physics Today* article, which appeared in 1970. This article produced the results I desired; from then on Everett could not be ignored.

The *Physics Today* article was deliberately written in a sensational style. I introduced terminology ("splitting," multiple "worlds," etc.) that some people were unable to accept and to which a number of people objected because, if nothing else, it lacked precision. I tried to remedy some of the article's defects in a series of lectures I was asked to give, not long after, at one of the Varenna summer schools. The article based on these lectures (MW, p. 167ff.) tries to define carefully what a "good" measurement is and to deal with the question of imperfect measurements and with the case of observables having continuous spectra, measurements of which are always imperfect and for which the "splits" into multiple worlds are never clean. I regard this latter article as my most careful statement of my views on the Everett interpretation.

Precisely because I did not wish to appear as the sole spokesman for Everett I conceived the plan of collecting in one volume the total world commentary, up to that time, on Everett. I was also convinced that Everett's "Relative State" paper could not have constituted a complete statement of his views. Wanting to give him the chance to express himself fully I began a search for additional material. First I secured a copy of his thesis from Princeton, only to find, as I should have expected from the first footnote in the RMP article itself (MW, bottom p. 141), that the article and the thesis are identical. It should be stressed that up to this time I had never met Everett. Although he did not exactly shun publicity, he certainly did not seek it and was, in any case, bored by the interpretation controversies. With John Wheeler's help, however, I was able to get Everett to send me a thick, faded, dog-eared manuscript entitled "The Theory of the Universal Wave Function." According to Everett this was his *Urwerk*, on which the *Reviews of Modern Physics* article was based. I therefore placed it first (MW, p. 3ff.) in the collection that Princeton University Press ultimately produced. Unfortunately Everett is now deceased, so an independent researcher cannot ask him to verify that this was his *Urwerk*. But one need not rely on my statement. Internal evidence already indicated that it antedates the RMP article. For example, in it Everett writes "Einstein hopes ..." (MW, p. 112, line 4), indicating that Einstein was still alive and that the work cannot have been written later than the spring of 1955.

On the question why this work, which is well written and brings Everett's views out more sharply than the RMP article, was never previously published, I can only speculate. I know that John Wheeler admires brevity and probably urged Everett to try to "sum up in a nutshell" the essential points of his new interpretation of quantum mechanics. It is also possible that Wheeler was reluctant to support a more blatant statement because it would mean setting himself into direct opposition to his hero, Niels Bohr. What is sure is that Wheeler long ago abandoned his support for Everett. What is equally sure is that if the *Urwerk* had been published Everett would not have been ignored for so long.

In 1972 I moved to The University of Texas and a few years later was enormously pleased to have Wheeler join our faculty from Princeton. One of the first things he did upon arrival was to persuade Everett to come to Austin and visit us for a few days, so that I was able to meet him at last. As he shook my hand he pulled out of his briefcase the original of my eleven-page letter and waived it under my nose. I assured him that I had forgotten neither the letter nor his reply. Everett was of small stature and a chain smoker. I was very saddened by his death not long after, as I would have loved to talk with him again many times. In none of our discussions in Austin was there any suggestion that our views on the interpretation of quantum mechanics were not in complete harmony, so I am a bit surprised to find that there are those who feel that I have not done justice to Everett. Of course, as I remarked earlier, Everett had long since ceased to think deeply about the interpretation question; his attention then was on the problem of artificial intelligence. But this is not wholly unrelated to the interpretation question: Was Everett a "realist"?

In his Urwerk one may note that he uses the terms "observer" and "automaton" interchangeably, with complete indifference (MW, top p. 7; p. 64, line 8ff.). It was clear from our Austin conversations that, for Everett, the possibility of artificial intelligence and machines possessing consciousness was obvious. The only questions concerned details. For Everett there could be no distinction between "consciousness" and "the contents of a memory bank." They were identical. I remember us getting quickly into philosophical questions such as free will. He said "You give me an operational definition of an automaton that has free will, and I will design a computer program that will simulate your automaton to any degree of detail that you may desire. Therefore free will – *your* definition of free will, mind you – exists." For him, it seemed, a computer program was virtually synonymous with the printout to which it ultimately led. They were interchangeable – equally "real." For him, whether we (i.e., the universe and all that is in it) have an independent existence or are merely solutions of some super differential equation is irrelevant. If there is an isomorphism between one and the other they are interchangeable. The words "one-one" and "isomorphism" already appear in Everett's Urwerk (MW, p. 109, line 3; bottom p. 133). Under an isomorphism between formalism and the "real" world, if something exists in the formalism then it "exists" in the "real" world. Does that make Everett a "realist"? In my opinion the views of both Everett and myself lie somewhere between realism and Platonic idealism. We both believe in the "reality" of the many worlds but we also believe that ultimately the abstract idea, theory, wave function, or ideal form behind it all is the true reality."

.6.2 Everett's Theory Viewed by John Wheeler

In his autobiography, written in collaboration with Kenneth Ford, Wheeler[52] recalls the 1957 Chapel Hill conference (see Sect. III.5), his contributions, and the contributions of his students (eight papers total) to the July 1957 issue of *Reviews of Modern Physics* . We note an article by Charles W. Misner based in part on his Ph.D. thesis, Hugh Everett III's Ph.D. thesis, and Wheeler's own assessment of Everett's "'Relative State' Formulation of Quantum Theory," among others.

[52] J. A. Wheeler with K. Ford *Geons: Black Holes, and Quantum Foam – A Life in Physics.* W. W. Norton & Company; (2000).

The following is a quote from Wheeler's autobiography.

"One very deep paper by my student Hugh Everett in this *Reviews* issue was so impenetrable that I was moved to publish next to it a short paper entitled "Assessment of Everett's 'Relative State' Formulation of Quantum Theory." Everett was an independent, intense, driven young man. When he brought me the draft of his thesis, I could sense its depth and see that he was grappling with some very basic problems, yet I found the draft barely comprehensible. I knew that if I had that much trouble with it, other faculty members on his committee would have even more trouble. They not only would find it incomprehensible; they might find it without merit. So Hugh and I worked long hours at night in my office to revise the draft. Even after that effort, I decided the thesis needed a companion piece, which I prepared for publication with his paper. My real intent was to make his thesis more digestible to his other committee members.

Everett's paper and my interpretation of it were concerned with the basics of quantum theory, only very loosely linked to relativity. The standard approach to quantum theory then (and now) assigns probabilities to the possible outcomes of quantum events. The actual outcome for a particular experiment is ascertained by a measurement that uses a "large", nonquantum detector – a piece of laboratory apparatus, for example, or the human eye. Only by replicating the same experiment many times can one confirm that the outcomes follow the predicted probabilities. And until the actual measurement is made, there is no way to know which among the possible outcomes will be realized.

A difficulty with this "Copenhagen interpretation", a difficulty that still deeply troubles me and many others, is that it splits the world in two: a quantum world, in which probabilities play themselves out, and a classical world, in which actual measurements are made. How can one clearly draw a line between the two? By how much must a quantum event be magnified to become a classical observation? When does probability give way to actuality?

Everett, in a tour de force, sought to get around these troubling questions by describing a totally quantum world in which there was no such thing as a classical observer, only quantum systems at all levels of size and complexity. Everett's "observer" is part of the quantum system, not standing apart from it. An oversimplified way to describe the outcome of his reasoning is to say that all of the things that might happen (with various probabilities) are in fact happening. Since there is no classical measuring apparatus in his formulation to determine which among the possible outcomes occur, one must assume that all the outcomes are occurring, but with no communication among them.

To see what this means, think of yourself driving down a road and coming to a fork. According to classical physics, you take one fork, and that's that. According to the conventional interpretation of quantum mechanics, you might take one fork or you might take the other, and which one you take will not be known until something happens to pin down your location, such as stopping at a gas station or restaurant, where some outside "observer" ascertains your location. There is something ghostly about even the conventional quantum interpretation, since it assumes that you travel "virtually" (as opposed to "really") down both roads at once, until it is established that you "really" traveled down a particular fork. According to the Everett interpretation, you go down both roads. If you stop later for gas on the left fork and someone observes you there and you are yourself aware of being there, that doesn't mean that there isn't another "you", uncoupled from the left-fork you, who stops to eat on the right fork, is observed by people there, and is aware of being there. Bryce DeWitt, my friend in Chapel Hill, chose to call the Everett interpretation the "many worlds" interpretation, and DeWitt's terminology is now common among physicists (although I don't like it). The idea has entered into the general public consciousness[53] through the idea of "parallel universes."Although I have coined catchy phrases myself to try to make an idea memorable, in this case I opted for a cautious, conservative term. "Many worlds" and "parallel universes" were more than I could swallow. I chose to call it the "relative state" formulation.

To me the important thing was not what analogies or fanciful visions one might spin out of Everett's work, but two basic questions: Does it offer any new insights? Does it predict outcomes of experiments that differ from outcomes predicted in the conventional quantum theory? The answer to the first question is emphatically yes. The answer to the second is emphatically no.

Should scientists care about new insights or different ways of looking at things if nothing new is predicted that can be measure? Yes, they should. We need to be always looking at what we already know in new ways. It's like an artist examining a piece of sculpture from every angle. The scientist, like the artist, might get a new idea or at least a deeper appreciation of what is already "known." There is no limit to depth of understanding. Different ways to describe the same set of equations can add insight. Maybe one way is clearly more economical, more "elegant" than another. Then we adopt that way. Maybe having two ways enriches our understanding by letting us examine the same domain of nature in two different ways. Maybe using one description will sometimes seem clearer at one time or for one purpose while using an alternate description will seem clearer at a different time or for a different purpose. In general relativity, for instance, it is sometimes easier to talk of the three-dimensional geometry of space evolving through time, and sometimes easier to talk of the four-dimensional geometry of spacetime that just "is." It is not a question of one description being right and the other wrong,

---

[53] In 1976, the Science Fiction magazine *Analog* printed a short story "Schrödinger's Bandits." Everett sent copies to his friends.

or even of one being better than the other. They are simply two ways to describe the same physics. For some applications, it may prove easier to use one approach, and for other applications the other approach.

What we have in Everett's work is a mind-stretching new way to look at quantum theory, one that triggers some very provocative thinking about the nature of the world even if it predicts no new experimental results. It may one day help germinate a better quantum theory or a better merger of quantum theory and relativity. Quantum theory has been around for most of the twentieth century, and its successes are legion. But the last word has not been written on it. I think about it every day."

III.7 Relativity, Groups, and Topology

Two major papers by Bryce DeWitt began as lecture notes at the Ecole d'Eté de Physique Théorique (Les Houches) in 1963 and 1983, respectively:

- "Dynamical Theory of Groups and Fields" [BD 23, 29], 262 pages.[54]
- "The Spacetime Approach to Quantum Field Theory" [BD 65, 66], 358 pages.

The Ecole de Physique at Les Houches had been created in 1951 by, then, Cécile Morette. The school was officially created on April 18 by the *Conseil de l'Université de Grenoble.* Cécile and Bryce were married on April 26 of the same year.

At the beginning, the School program included a basic course on Quantum Mechanics and a basic course in Statistical Physics each year; these courses were badly needed in most countries outside of the USA, Canada, and the UK. Cécile invited Bryce to give the Quantum Mechanics course in 1953. His lecture notes (517 typewritten pages), typed on acid paper and reproduced by stencils for distribution to the participants[55], have been scanned thanks to Molly White, the librarian for Physics, Mathematics, and Astronomy at the University of Texas. They are available at <http://repositories.lib.utexas.edu/handle/2152/19>.

By 1958 the need for basic introductory courses was not as acute as in the early fifties and sessions were organized around a theme of current interest. Given the length of the sessions and the length of the courses, the

[54] The handwritten solutions to the problems in *Dynamical Theory of Groups and Fields* have been scanned. They are available at http://repositories.lib.utexas.edu/. The originals have been deposited in the archives of the UT Center for American History; *see* Sect. V.4.
[55] Distributed within 24 hours of the lecture according to the rule of the School for a couple of decades!

lecturers presented recent developments in a larger context, including their limitations and their potential. The themes in 1963 and 1983 were "Relativity, Groups, and Topology."[56] The 1963 session was organized by Bryce and Cécile DeWitt, the 1983 session by Bryce DeWitt and Raymond Stora.

The 1963 DeWitt lecture notes have been reproduced in a book [BD 29].

Preface to the 1963 Les Houches Lecture Notes

"This book is based on a series of lectures which I gave at the Les Houches Summer School in 1963. It was first published as part of a larger volume, *Relativity, Groups, and Topology* (Gordon and Breach, 1964) which contains all of the lectures given at Les Houches that summer. A few trivial mistakes in the original have been corrected in the present volume, and a chapter has been added at the end which sketches in barest outline a theory of the higher order radiative corrections for non-Abelian gauge fields, a problem that was left dangling in 1963.

So far not a shred of experimental evidence exists that fields possessing non-Abelian infinite dimensional invariance groups play any role in physics at the quantum level. And yet motivation for studying such fields in a quantum context is not entirely lacking. It is only by asking quantum questions that one has in recent years been led, for example, to discover some of the deeper properties of the classical gravitational field. Furthermore, some of the fundamental concepts of field theory stand fully revealed only in a non-Abelian setting.

The extension of the quantum theory to fields possessing non-Abelian infinite dimensional invariance groups is harder than one might expect solely on the basis of experience with quantum electrodynamics and its Abelian gauge group. The difficulties seem to be mainly technical, but their resolution is not without interest and reveals a most intricate interplay between group and field which is absent in the purely classical theory. The relevant and rather recondite details are to be found in Chapters 23 and 25.

Aside from closing this previously existing theoretical gap, the present volume has the more modest aim of trying to represent field theory as a worthy object of affection. Field theory is nowadays too often held up to scorn by pioneers working on the raw frontiers of physics. This scorn stems from a misconception of field theory's proper role. She is not a robust mate ready to pitch in and lend a helping hand. She is a haunting mistress, refined, and much too beautiful for hard work. She is at her best in formal dress, and is thus displayed in this book, where rigor will be found to be absolutely absent.

[56] C. M. DeWitt and B. S. DeWitt (eds) "*Relativity, Groups, and Topology*" 1963 Les Houches Lectures. Gordon and Breach, New York (1964) 929 pp; Bryce DeWitt and Raymond Stora (eds.) "Relativity, Groups, and Topology II". North Holland, Amsterdam (1984), 1323 pp.

I have eschewed the canonical approach throughout. The invariance group is always dealt with in a manifestly covariant manner. Contact with quantum theory is made through the theory of measurement and the laws of propagation of small disturbances. The S-matrix is introduced in the LSZ manner and special attention is devoted to the modifications which the presence of an invariance group imposes upon the usual discussion of asymptotic fields, Feynman diagrams, and external line wave functions. No calculations of physical processes have been included, although the derivations are in several cases carried right up to the point where explicit calculations could begin in a straightforward (but tedious) manner. This book is therefore not in any sense a textbook. However, if a textbook containing traditional material were to be written in the same style it could avoid the traditional shock transition zone between those chapters which present the abstract theory and those which list the Feynman rules.

One calculation which is included and which is perhaps worth mentioning is a demonstration in Chapter 24 that renormalization can be carried out directly in configuration space if intelligent use is made of the structural information about Green's functions obtained in earlier chapters. This possibility greatly alleviates the difficulty of demonstrating the covariance of renormalization procedures, which is especially acute when space-time is curved.

Several chapters are devoted to purely group theoretical matters. Their purpose is to provide a necessary vocabulary and to reveal the structure underlying much of field theory. Included also are over a hundred exercises whose purpose is to amplify the text.[54]

I wish to express my gratitude for my wife who, in insulating me from children and colleagues at the appropriate moments, made the writing of this book bearable as well as terminable."

The book was later translated into Russian and remains a reference work. It includes a new preface dated 1985 and an introduction by G. A. Vilkovisky dated 1986. Bryce's wife and children having objected to the acknowledgment in the English version, it was deleted in the Russian version and replaced with the following acknowledgment.[57]

я хочу выразить благодарность директору и Совету попечителей Летней школы в Лезуш за предоставленную мне возможность
выступить с этими лекциями перед избранной аудиторией студентов

[57] "I wish to thank the director and board of trustees of Les Houches summer school for the opportunity of giving these lectures to an elite audience".

Introduction to the 1983 Les Houches Lecture Notes

"Twenty years ago, at the Les Houches session bearing the same title as the present one, I began my lectures with the words 'A chief goal of these lectures is to develop the framework within which the quantization of fields possessing infinite-dimensional invariance groups may be carried out in a manifestly covariant fashion. The requirement of manifest covariance means that we shall adopt an over-all spacetime view from the outside and ignore canonical formulations.

The goal today remains the same. I have been astonished to see over the intervening years how physicists, when expounding the fundamentals of field theory, have failed to apply the lessons that relativity theory taught them early in this century. Although they usually carry out their calculations in a covariant way, in deriving their calculational rules they seem unable to wean themselves from canonical methods and hamiltonians, which are tied to the cumbersome (3+1)-dimensional baggage of conjugate momenta, bigger-than-physical Hilbert spaces, and constraints. There seems to be a feeling that only canonical methods are 'safe'; only they guarantee unitarity. This is a pity because such a belief is incorrect and it makes the foundations of field theory unnecessarily complicated. One of the unfortunate results of this belief is that physicists, over the years, have almost totally neglected the beautiful covariant replacement for the canonical Poisson bracket that Peierls invented in 1952[58]. In my 1963 lectures the Peierls bracket formed a cornerstone of the theory. I champion it once again here (section 4).

I do not mean to imply that a (3+1)-dimensional language is not sometimes useful. Indeed the dynamical system may have a certain property, or be in a certain state, that renders such a language appropriate, for example when there exists a global timelike Killing vector field. But it need not be the standard canonical language."

Two major advances occurred during the twenty year interval between the two Les Houches sessions with the same theme, "Relativity, Groups, and Topology."

- The need for ghosts for the functional integrals used in quantized gravitation. For a discussion of ghosts see Sect. I.2, "The Geometry of Gauge Fields." The story of ghosts is a feature of early quantum gravity research.
- The 1983 lecture notes use superanalysis throughout, and include an appendix that is a compendium of superanalysis. This branch of mathematics is a tool well adapted to a systematic handling of parity and its

[58] R. E. Peierls "The commutation laws of relativistic field theory", *Proc. Roy. Soc. (London)* **A214**, 143–157 (1952).

applications in all branches of algebra, analysis, geometry, and topology. If an object A has a parity $\tilde{A} \in \{0, 1\}$ it is called even if $\tilde{A} = 0$ and odd if $\tilde{A} = 1$.

Given the importance of parity and the contributions of DeWitt to supermanifolds, the subject is presented separately in Sect. III.9.

III.8 Quantum Field Theory in Curved Spacetime[59]

"Bryce's Physics Report made a bridge between 'quantum' and 'gravity' based on firm physics. This paper is very important. In those days, it was like a fresh air cleaning brains."[60]

Indeed, more than thirty years later, this report remains a landmark for quantum gravity with results that are still of current interest. The introduction to the report reads as follows:

"The existence of the Poincaré group as a local symmetry group for spacetime has been enormously important to particle physicists in helping them sort out their ideas and to construct formalisms for describing experimental facts – formalisms that run the gamut from pure phenomenology through dispersion theory to axiomatic field theory. In fact, students are taught nowadays that elementary particles simply *are* certain representations of the Poincaré group.

An addiction of any kind ultimately extracts a penalty from the addict. Physicists learned this lesson well in the early decades of this century. Most of us are aware that Quantum Field Theory cannot in the end be based on the Poincaré group. What is needed is a theory – or at least a framework – that respects the full general covariance of Einstein's view of spacetime as a riemannian manifold.

It is not my purpose here to present such a theory; it does not yet exist, at least as a coherent discipline. What I shall do is describe several distinct but related examples of physical processes that involve the manifold structure of spacetime in an essential way and that show some of the important elements that must go into such a theory. These examples are chosen both for their pedagogical value and for their current interest, and I hope that they will convince the reader not only that a coherent theory can ultimately be built but that it will also be extremely beautiful.

[59] [BD 51]: *Physics Reports* **19c**, 295–357 (1975). Translated into Russian for the series Cherniye Diri: Novosti Fundamentalnoi Fisike. Mir, Moscow (1978).
[60] G. A. Vilkovisky, personal communication.

The core of any theory of interacting fields is the set of currents that describe the interaction.

The currents of general relativity theory are the components of the stress tensor. A fundamental task – I might even say *the* main problem – in developing a Quantum Field Theory in curved spacetime is to understand the stress tensor. The stress tensor, like any current, is formally a bilinear product of operator-valued distributions (the field operators) and hence is meaningless. The problem is to give it meaning, by some subtraction process.

A subtraction, or regularization, procedure conventionally makes use of the vacuum state. Particle physicists know what the vacuum is: It is (modulo symmetry breaking degeneracies) the trivial representation of the Poincaré group. General relativists are not so lucky. In the absence of geometrical symmetries they have many 'vacua' to choose from."

The two concrete examples chosen by DeWitt are

The Casimir effect[61]

"This well known effect, predicted and popularized by Casimir and experimentally confirmed in the Philips laboratories, has at first sight nothing to do with curvature: Two extremely clean, neutral, parallel, microflat conducting surfaces, in a vacuum environment, attract one another by a very weak force that varies inversely as the fourth power of the distance between them.

However, just as curvature can be regarded as a cluttering up of spacetime with bumps, so can the Casimir apparatus be regarded as a cluttering-up of spacetime with neutral conductors. Although the effect was first computed as a kind of Van der Waals force, because the force turns out to be independent of the molecular details of the conductors Casimir quickly recognized that it could be computed as a problem in vacuum energy, and that is the way it is computed in the classroom today.

[61] H. B. G. Casimir "On the attraction between two perfectly conducting plates", *Proc. Kon. Nod. Akad, Wetenschap.* **51**, 793–795 (1948), *see also* T. H. Boyer "Quantum zero-point energy and long-range forces", *Ann. Phys. (N.Y.)* **56**, 474–503 (1970), and Bryce DeWitt "The Casimir Effect in Field Theory", in: A. Sarlemijn and M. J. Sparnaay (eds.) *Physics in the Making – Essays in Honor of H. B. G. Casimir.* North Holland, Amsterdam, pp 247–272 (1996).
For a general introduction to the Casimir effect, one can read: A. Gambassi, C. Hertlein, L. Helden, S. Dietrich, C. Bechinger "The Critical Casimir Effect, universal fluctuation – induced forces at work", *Europhysics News* **40**, 18–22 (2009).

It is true that the tiny energy involved is too small by many orders of magnitude to produce a gravitational field that anybody is going to detect, but one can easily construct Gedankenexperimente in which the laws of conservation of energy is violated unless this energy is included in the source of the gravitational field. Relativists should note that the energy density involved is negative, and hence the stress tensor violates the classical energy theorems so crucial to blackhole theory[62]. Everybody should note that the Casimir energy is a pure vacuum energy; no real particles are involved, only virtual ones. And experiments tell us that we have to take it seriously."

Accelerating Conductors

The issue is particle production by moving boundaries[63]. In DeWitt's words:

"The Casimir effect may be called a pre-curvature effect of manifold structure. Before going on to discuss true curvature effects let me follow Einstein's example by first discussing effects caused by acceleration. In applying the thermodynamical law (34) to the Casimir vacuum stress I required that the conductors be moved slowly. If I were to accelerate them appreciably they would emit photons, and the entropy in the slab region would be increased. It may seem surprising at first that by accelerating a *neutral* conductor one can produce photons, but then one quickly remembers that the surface layers of a real conductor carry currents. The free electrons near the surface react to the quantum fluctuations of the electromagnetic field just as they do to a classical field and produce currents of just the required amount to guarantee the standard boundary conditions. Because the boundary conditions suffice to determine the physics outside the conductors one need not refer to the currents, as such, at all."

To see how this works, the reader is referred to the technical pages of *Physics Reports*, pp 308–312, which are beyond the scope of this book.

Quantum Field Theory in curved spacetime is introduced by a study of the Kerr black hole "because it illustrates well a great new range of problems." Sections 4 (The Kerr black hole), 5 (Exploding black holes), and 6 (The divergences) should be required reading in any course on Quantum Gravity.

The report ends with the future outlook as it appeared in 1975. The outlook spelled out concretely is rather glum.

---

[62] "The negativity of the energy appears to be a function of conductor geometry. Boyer [loc. cit. in *Physics Reports* 4] and Davies [loc. cit. in *Physics Reports* 17] have shown that the vacuum energy inside a conducting sphere is positive."

[63] In a note added in proof, DeWitt refers the readers to the "valuable" article by G. T. Moore "Quantum Theory of the Electromagnetic Field in a Variable-Length One-Dimensional Cavity", *J. Math Phys.* **11**, 2679–2691 (1970).

.9 Supermanifolds

Complex numbers generalize real numbers. Supernumbers generalize complex numbers. They are useful to codify parity.

Parity is ubiquitous. Supernumbers are numbers that have parity. Algebra and analysis that use supernumbers are called "Grassmann Algebra" and "Grassmann analysis." They are tools well adapted for handling systematically parity and its implications in all branches of algebra, analysis, geometry, and topology.

The parity rule that describes the behavior of a product under exchanges of its two factors is sometimes called Koszul's parity rule. It states that '*Whenever you interchange two factors of parity 1, you get a minus sign.*' Formally the rule defines graded commutative products

$$AB = (-1)^{\tilde{A}\tilde{B}} BA$$

where $\tilde{A} \in \{0, 1\}$ denotes the parity of A. Objects with parity zero are called *even*, and objects with parity one *odd*. The rule also defines graded anticommutative products. For instance,

$$A \wedge B = -(-1)^{\tilde{A}\tilde{B}} B \wedge A.$$

The co-existence of bosons and fermions tells physicists that Grassmann analysis is the tool, *par excellence* of Quantum Field Theory. It was introduced in Quantum Field Theory in 1965 by F. A. Berezin and developed in 1975, and further in 1977 in collaboration with M. S. Marinov.

The 1954 unpublished marathon (eleven hours: seven in one day, and four in the next) lecture of Schwinger motivated Bryce DeWitt to treat bosons and fermions simultaneously. What had been a rudimentary application of the supernumber of a Grassmann algebra became in his hands an extensively developed superanalysis.

The prefaces to the first and second editions of *Supermanifolds*[64] read as follows:

Preface to the first edition

"This book is an outgrowth of a book on quantum gravity that the author started to write nine years ago in collaboration with Christopher Isham. It began as an Appendix to the quantum gravity book, but subsequent developments modified the original plan. Firstly, new results in quantum gravity, particularly in supergravity and in the applications of topology to quantum

64 [BD 67, 91].

Fig. 10 Bryce and Christopher J. Isham. *Supermanifolds* began as an appendix to a book that Bryce and Chris were going to write together. The handwritten table of contents of the book combining sections in B. DeWitt's and C. Isham's handwritings has been deposited at the Archives of the UT Center for American History

field theory, appeared so rapidly that the timing of the collaborative volume became inopportune. Secondly, the theory of supermanifolds had so many loose ends which needed to be dealt with that the original Appendix grew beyond reasonable size limits and turned into a book in its own right.

A previous generation of theoretical physicists could function adequately with a knowledge of the theory of ordinary manifolds and ordinary Lie groups. With the discovery of Bose-Fermi supersymmetry all this changed. Nowadays the theorist must know about supermanifolds and super Lie groups. The purpose of the present volume is to provide him with an easily accessible account of these mathematical structures. Mathematicians will find much of this book incomplete and expressed in language that they have nowadays passed beyond, but it is probably pitched about right for the average physicist. It still has something of the character of an Appendix in its lack of any account of how it relates to supergravity and other locally supersymmetric theories. For a time it was to have appeared as Volume I of a two-volume work on supermanifolds and supersymmetry written in collaboration with Peter van Nieuwenhuizen and Peter West. However, delays caused by new developments in supergravity theory, particularly in higher dimensional Kaluza-Klein versions of supergravity theory, rendered this linkage impractical. The second volume will ultimately appear (in the same Cambridge University Press series), but rather than delay the first volume further, the decision was made to publish the two as separate books. While waiting for the second book to appear, the reader of the present volume who wishes to establish linkages to physics will have to content himself with studying the elementary applications of supermanifold theory selected in chapter 5 of this volume and with reading the already vast literature on supersymmetric theories.

This author wishes to express his gratitude to the following for their support at various times during the writing of this book: The Warden and Fellows of All Souls College, Oxford, where the book was begun, the John Simon Guggenheim Foundation, the United States National Science Foundation, the North Atlantic Treaty Organization and The University of Texas."

Preface to the second edition

"At the end of the fifth and last chapter of the first edition the author wrote that if the book were ever to be revised it would include an account of the beautiful work of E. Witten[65] and of L. Alvarez-Gaumé[66] on supersymmetry, Morse theory, and the Atiyah-Singer index theorem. Chapter 6 of this revised edition is a partial fulfillment of that promise. The aim of chapter 6, like that of chapter 5, is almost exclusively pedagogical. Unlike chapter 5, however, chapter 6 deals with nontrivial supermanifolds, and the author discovered that there are numerous fine points in the theory of the Feynman functional integral for such supermanifolds that are not adequately covered in the literature, even on a formal level. To be pedagogically helpful the book *has* to deal with these issues, given the fact that, despite the essential role it plays in chapter 6, the functional integral is used in a formal way rather than as a rigorous tool. This has meant that, in order to keep the reader's confidence, the author has had expend a large part of his effort on the functional integral itself and hence could include only a little of the flavor of the index theorem, as it touches the Euler-Poincaré characteristic. The effort to display the internal consistency of the functional integral formalism has nevertheless been useful in that it presents a challenge to the student to attempt what must surely be possible, namely, to establish the functional integral at last on a fully rigorous basis for both bosonic and fermionic systems."

[65] E. Witten "Supersymmetry and Morse Theory", *J. Diff. Geom.* **17**, 661–692 (1982).
[66] Luis Alvarez-Gaumé "Supersymmetry and Index Theory," A set of lectures given at the 1984 NATO Summer School in Bonn, Germany.

Progress towards this challenge can be found in the Cartier, DeWitt-Morette book[67].

A *technical caveat* is mentioned here because Grassmann algebra is not familiar territory, and looks cumbersome at first sight. It is even more confusing since different authors use different conventions. Bryce DeWitt defines complex conjugation of the product of two supernumbers z_a and z_b by

$$(z_a z_b)^* = z_b^* z_a^* \quad \text{(hermitian conjugation rule)}$$

Pierre Cartier and Cécile DeWitt-Morette define the complex conjugation of a product by

$$(z_a z_b)^* = z_a^* z_b^* \quad \text{(complex conjugation rule)}$$

With the hermitian conjugation rule the product of two *real* supernumbers is purely *imaginary*; with the complex conjugation rule it is *real* (as it should be!). By the time the advantages of the complex conjugation rule were brought to the attention of Bryce DeWitt, he had already fully developed his book: *Supermanifolds*. A pity, but as written on the jacket of the second edition of the book: "*Supermanifolds* is destined to become the standard work for all serious study of super-symmetric theories in physics. (*Nature*)".

III.10 The Global Approach to Quantum Field Theory

The voice of Bryce DeWitt can be again heard in the prefaces of his books. His last book "The Global Approach to Quantum Field Theory," 2 volumes totaling 1042 pages, was first published in 2003, and reprinted with minor corrections in 2004.

"There exists an anomaly today in the pedagogy of physics. When expounding the fundamentals of Quantum Field Theory physicists almost universally fail to apply the lessons that relativity theory taught them early in the twentieth century. Although they usually carry out their calculations in a covariant way, in deriving their calculational rules they seem unable to wean themselves from canonical methods and hamiltonians, which are holdovers from the nineteenth century and are tied to the cumbersome (3+1)-dimensional baggage of conjugate momenta, bigger-than-physical Hilbert spaces, and constraints. There seems to be a feeling that only canon-

[67] P. Cartier and C. DeWitt-Morette, *Functional Integration, Actions and Symmetries* (Cambridge University Press, 2006).

ical methods are "safe"; only they guarantee unitarity. This is a pity because such a belief is wrong, and it makes the foundations of field theory unnecessarily cluttered. One of the unfortunate results of this belief is that physicists, over the years, have almost totally neglected the beautiful covariant replacement for the canonical Poisson bracket that Peierls invented in 1952.[68]

Historically Quantum Field Theory was built on classical field theory, just as quantum mechanics was built on classical mechanics. To the author this still seems a good approach, provided the links between the classical and quantum theories are displayed elegantly. One of the purposes of this book is to provide the reader with a fully relativistic, global view of spacetime and its dynamics ab initio. The Peierls bracket will play a central role in the development. Its definition is intimately tied to the theory of measurement, as is revealed in a classic paper by Bohr and Rosenfeld[69] where it already appears in rudimentary form.

It also carries one easily, indeed irresistibly, forward to the heuristic quantization rules embodied in the Schwinger variational principle and the Feynman sum over histories.

The global approach does not prevent one from appreciating the traditional canonical theory. In appropriate situations, canonical methods are both highly useful and strikingly beautiful. But it is generally easier to descend to them from the global vantage point that to climb in the reverse direction. They are always accessible and can be brought into play whenever it is convenient to do so. It is often convenient when the specifically (3+1)-dimensional character of spacetime is of primary importance, for example when there exists a global timelike Killing vector field, or when thermal properties are under study.

One cannot do without (3+1)-dimensional assumptions. Although space and time together comprise a single geometrical entity, individually they are distinct. Basic to the whole of Quantum Field Theory is the assumption that spacetime, which we shall denote by M (for manifold), has the topological structure

$$M = \mathbb{R} \times \Sigma$$

where $\mathbb{R}$ is the real line and Σ is some connected three-dimensional manifold, compact or noncompact. More precisely, spacetime will be assumed to be endowed with a hyperbolic metric which admits a foliation of spacetime

---

[68] "R. E. Peierls "The commutation laws of relativistic field theory," *Roy. Soc. (London)* **A214**, 143–157 (1952)."

[69] "N. Bohr and L. Rosenfeld, Kgl. *Danske Videnskab. Selskab, Mat.-fys. Med.* **12**(8) (1933)." *see* footnote 36 in Sect. III.4.

into spacelike sections, each being a complete Cauchy hypersurface (i.e., a hypersurface on which initial data can be completely specified) and a topological copy of Σ.[70]

It will be convenient to generalize from four to n dimensions, so that the spatial section Σ has $n - 1$ dimensions. n will sometimes be formally extended into the complex plane (dimensional regularization). It will sometimes be equal to 1. In this case Σ reduces to a point and the dynamical theory becomes that of ordinary quantum mechanics...

A certain sophistication will be required of the reader, who will be assumed to be familiar with differentiable manifolds and Lie groups, and with at least the rudiments of the theory of fibre bundles and of supermanifolds. For readers whose knowledge of these subjects needs to be refreshed the following books are recommended: *Analysis, Manifolds and Physics Part I (Revised Edition 1989)*; *Part II (Revised Edition 2000)* by Y. Choquet-Bruhat and C. DeWitt-Morette with M. Dillard-Bleick (North Holland, Amsterdam, 1989) and *Supermanifolds* (Second Edition) by B. DeWitt (Cambridge University Press, 1991). For readers whose needs are relatively minor and who merely need to be refreshed on such things as *a*-numbers (anticommuting supernumbers), *c*-numbers (commuting supernumbers), and Berezin integration, the Appendices at the end of the book will suffice.

The book is in no sense a reference book on Quantum Field Theory and its application to particle physics. The selection of topics is idiosyncratic. It has, however, the aim of showing how the whole structure hangs together. Because of the book's emphasis on the Peierls bracket and the global point of view, a number of topics, notably the theory of conservation laws and their relation to background fields, are developed by methods that will be unfamiliar to many readers. Emphasis is also given to the astonishingly varied role played by the measure in the functional integral.

Each of the book's 35 chapters is followed by a paragraph of Comments together with a short list of references (by no means complete). The chapters themselves are grouped into seven parts: I. Classical Dynamical Theory; II. The Heuristic Road to Quantization. The Quantum Formalism and Its Interpretation; III. Evaluation and Approximation of Feynman Functional Integrals; IV. Linear Systems; V. Nonlinear Fields; VI. Tools for Quantum Field Theory. Applications; VII. Special Topics. Supplementing these is an eighth part containing 25 elementary examples illustrating basic results or procedures introduced in the preceding chapters. These examples should be read in parallel with the other parts of the book...
March 2002 B. DeW."

---

[70] "When the gravitational field is quantized the expectation values of the metric in all physically acceptable states will be assumed to be endowed with these properties (or with some physically motivated natural extension thereof)."

The book requires a sustained effort from the reader. Is it worth it? What does the book offer that cannot be found elsewhere? The following review by Alvarez-Gaumé[71] answers these questions:

"It is difficult to describe or even summarize the huge amount of information contained in this two-volume set. Quantum Field Theory (QFT) is the more basic language to express the fundamental laws of nature. It is a difficult language to learn, not only because of its technical intricacies but also because it contains so many conceptual riddles, even more so when the theory is considered in the presence of a gravitational background. The applied field theory techniques to be used in concrete computations of cross-sections and decay rates are scarce in this book, probably because they are adequately explained in many other texts. The driving force of these volumes is to provide, from the beginning, a manifestly relativistic invariant construction of QFT.

Early in the book we come across objects such as Jacobi fields, Peierls brackets (as a replacement of Poisson brackets), the measurement problem, Schwinger's variational principle and the Feynman path integral, which form the basis of many things to come. One advantage of the global approach is that it can be formulated in the presence of gravitational fields. There are various loose ends in regular expositions of QFT that are clearly tied in the book, and one can find plenty of jewels throughout: for instance a thorough analysis of the measurement problem in quantum mechanics and QFT, something that is hard to find elsewhere. The treatment of symmetries is rather unique. DeWitt introduces local (gauge) symmetries early on; global symmetries follow at the end as a residue or bonus. This is a very modern point of view that is spelled out fully in the book. In the Standard Model, for example, the global symmetry (B-L, baryon minus lepton number) appears only after we consider the most general renormalizable lagrangian consistent with the underlying gauge symmetries. In most modern approaches to the unification of fundamental forces, global symmetries are quite accidental. String theory is an extreme example where all symmetries are related to gauge symmetries.

There are many difficult and elaborate technical areas of QFT that are very well explained in the book, such as heat kernel expansions, quantization of gauge theories, quantization in the presence of gravity and so on. There are also some conceptually difficult and profound questions that DeWitt addresses head on with authority and clarity, including the measurement problem mentioned previously and the Everett interpretation of quantum mechanics and its implications in quantum cosmology. There is also a cogent and impressive study of QFT in the presence of black holes, their

---

[71] *Cern Courier*, April 1 (2004); also recommended are the reviews by Stanley Deser in: *Physics Today*, March (2004, pp 77–78) "Quantum Reality, Ghosts, and more: A Pioneer Looks at QFT" and by Antoine Folacci and Bruce Jensen in: *Journal of Physics A* **36**, 12346–12347 (2003).

Hawking emission, the final-state problem for quantum black holes and a long etcetera.

The book's presentation is very impressive. Conceptual problems are elegantly exhibited and there is an inner coherent logic of exposition that could only come from someone who had long and deeply reflected on the subject, and made important contributions to it. It should be said, however, that the book is not for the faint hearted.[72] The level is consistently high throughout its 1042 pages. Nonetheless it does provide a deep, uncompromising review of the subject, with both its bright and dark sides clearly exposed. One can read towards the end of the preface: "The book is in no sense a reference book in Quantum Field Theory and its applications to particle physics...". I agree with the second statement but strongly disagree with the first."

One important feature of the book is that DeWitt does not limit himself to first approximations, be they linear or one-loop approximations. He does not shy away from the long, often uninspiring calculations of the second approximation. They often reveal obstacles not visible in the first approximation, and sometimes they pave the way to exact results.

The first approximation is similar to a plane in holding pattern, the scenery is tantalizing. The second approximation is similar to landing, the obstacles are real. The obstacles visible in second approximations cannot be ignored – even in more modern contexts.

III.11 The Pursuit Goes On

To my regret, I am not qualified to assess recent work in Quantum Gravity.[73] I shall limit myself to two items that I have come across in recent years:

- G. A. Vilkovisky's calculations of the backreaction of the Hawking radiation.
- A workshop held at the *Institut Henri Poincaré* (May 26–28, 2008) on "Gravitational Scattering, Black Holes, and the Information Paradox."

[72] The shortcoming of the otherwise-powerful condensed notation is that indices do double duty as coordinates in the domain of a function, as well as in its range. Throughout the book $h = 1, c = 1$. Inserting h and c occasionally would help the reader identify the physical dimensions of a term.

[73] A sample of current research can be found in the Les Houches Session LXXXVII lecture notes, "String theory and the real world from particle physics to astrophysics", C. Bachas, L. Baulieu, M. Douglas, E. Kiritsis, E. Rabinovici, P. Vanhove, P. Windey, L.F. Cugliandolo (eds.) Les Houches Session LXXXVII, Elsevier, Amsterdam (2008). For an overview survey *see*, for instance: I.K. Klebanov and J. M. Maldacena "Solving Quantum Field Theories via Curved Spacetimes" *Physics Today* Jan. 2009, p 28-33.

The Backreaction

In 2004 Gregory A. Vilkovisky was tackling the difficult calculation of the backreaction of the Hawking evaporation of black holes. He was hoping to have some results in time to share them with Bryce DeWitt, his long time friend.[74] It would have been a fitting gift to Bryce who always wanted to see a theory pushed to its logical conclusion.

Fig. 11 Bryce and Gregory Vilkovisky (Grisha) – The pursuit goes on... This last section includes some work by Grisha, a long time friend of Bryce's, completed after Bryce's death

[74] G. A. Vilkovisky "The gospel according to DeWitt" in: S. M. Christensen (ed.) *Quantum Gravity: Essays in honor of the 60th birthday of Bryce S. DeWitt*. Adam Hilgers, Bristol (1984).

Calculations of backreactions are basic to Quantum Field Theory. In four fundamental papers[75] Vilkovisky worked out the semi-classical collapse of a macroscopic spherically symmetric, uncharged body. The calculation is carried out in 1+3 dimensions. As a black hole forms and evolves its Hawking process[76] decreases its initial mass by about 10% and creates a vacuum-induced "charge", a "matter charge".

The collapse finishes with a true stable black hole of mass microscopically exceeding the vacuum-induced charge. The vacuum-induced charge corresponds to a stress-energy tensor of the type associated with a long-range field.

This quantitative result has been criticized by some who, on the basis of qualitative arguments, judge it highly improbable. Nevertheless the fact that the collapse of blackholes is limited by vacuum-induced charges has many consequences worth investigating.

A 2008 Workshop

The 2008 workshop was a delightful experience for me. Most of the participants were old friends of mine, some of them dating from the early fifties. None of the presentations I attended used powerpoints or transparencies. They were all "blackboard and chalk" and understandable. The program consisted of six questions at the core of Quantum Gravity. These questions can serve as the last page of my scrapbook of the early years of Quantum Gravity:

- What is the blackhole information paradox/problem/puzzle/question? Convener: T. Jacobson.
- Is string theory providing a statistical-mechanics interpretation of blackhole thermodynamics? Convener: J. Maldacena.
- Do quantum blackhole microstates have something to do with a classical geometry? Convener: D. Amati.
- Can the AdS/CFT correspondence teach us how to solve the information paradox? Convener: J. Polchinski.
- What can we learn from the study of transplanckian-energy collisions? Convener: G. Veneziano.
- Blackhole and evaporation workshop: Did we gain any new information?

---

[75] G. A. Vilkovisky "Kinematics of evaporating black holes" *Ann. Phys.* **321**, 2717–2756 (2006); G. A. Vilkovisky "Radiation equations for black holes" *Phys. Lett. B* **634**, 456–464 (2006); G. A. Vilkovisky "Backreaction of the Hawking radiation" *Phys. Lett. B* **638**, 523–525 (2006). A 2008 paper completes these three papers to the limit of validity of semiclassical theory: G.A. Vilkovisky "Post-radiation evolution of black holes" *Phys. Rev. D* **77**, 124043 (2008).

[76] Also known only at the semiclassical approximation level.

I cannot list all the physicists who, in one way or another, have contributed to the substance of this book. Since I am concluding this book with the 2008 workshop, I shall arbitrarily use a few names from the list of its participants to symbolically represent many others:

Daniele Amati, Costas Bachas, Curtis Callan, Marcello Ciafaloni, Thibault Damour, Bernard de Wit, Francois Englert, Pierre Fayet, Steven Giddings, Joao Gomes, Jim Hartle, Stephen Hawking, Gerard 't Hooft, Gary Horowitz, Stephen Hsu, Jean Iliopoulos, Ted Jacobson, Bernard Julia, Elias Kiritsis, Juan Maldacena, Joseph Polchinski, Eliezer Rabinovici, Andrew Strominger, Gabriele Veneziano, and Erik Verlinde; ... with my regrets and apologies to the many friends whose names do not appear on the 2008 workshop list.

A Few Comments from Bryce DeWitt

7.1 Amateurs, Crackpots, Professionals, and Gravitation

Professional letters will be stored in the DeWitts' archives at The University of Texas at Austin Center for American History. The following letter has been chosen for inclusion (names withheld) in these memoirs because it talks about non-specialists making physics theories on topics that strike their fancies, but who do not have a solid technical knowledge of the subject. This letter may be of interest to a larger public than other letters.

DeWitt became reacquainted with Jean deValpine, a former Harvard classmate, when deValpine's son John was a student at The University of Texas at Austin, and took a Plan II honors Seminar from Cécile (Fall 1987).

"Mr. Jean E. de Valpine
Memorial Drive Trust
Acorn Park
Cambridge, Massachusetts 02140

Dear Jean,
Many thanks for the material on the Nautilus Fund and for S.'s book.

It is a sad fact of life that of all physical theories (statistical mechanics, electromagnetic theory, condensed matter physics, nuclear physics, the quantum theory, etc.) only Einstein's theory of gravity and spacetime has captured the public imagination. It is also the only theory that gets regularly attacked – by crackpots as well as by potentially good physicists. Whenever I write a popular article on gravity (for example, my Scientific American article of a couple of years ago) I know and dread the fact that I am going to get at least a half a dozen crank letters. Do I owe it to the taxpayers to answer these letters? Or do I simply ignore them and reinforce the image of my own arrogance as a member of an elite having vested interests?

C. DeWitt-Morette, *The Pursuit of Quantum Gravity: Memoirs of Bryce DeWitt from 1946 to 2004*, DOI 10.1007/978-3-642-14270-3_5, © Springer-Verlag Berlin Heidelberg 2011

Authors of articles on evolution face similar problems, but in some ways physicists specializing in relativity have it worse, for we not only have to face the cranks from outside academia but also from a fringe group inside. The latter consists mainly of frustrated and bitter individuals who have usually had a paper containing a pet idea of theirs repeatedly rejected for publication and who have resorted to the following strategy: They send copies of their papers to dozens of specialists in the field (I have received many such) with a cover letter stressing the importance of their work, or of a proposed experiment, and asking for a detailed analysis of their ideas. What does one do with such a letter? Reading papers takes a lot of time and effort. The older one gets (and the more entrenched one gets in the "in" group) the more one feels pressed for time to do one's own research. Usually the author's rough idea (although not the details) has already occurred to one years ago and been dismissed for a variety of reasons, some of which may be no more than intuition.

Theoretical physics is less a science than an art and what physicists esteem in a theory is very subjective, having to do with elegance and predictive power. And yet there is surprising unanimity on what constitutes "good" physics. It is not a question of maintaining the status quo. Upsetting the applecart is the way Nobel prizes are won, and there are plenty of youngsters hoping to do just that. Progress demands not just new ideas but exciting new ideas.

I do not have time to take you through the purely physics part of S.'s book and point out the places where not only his judgement is unsound but he is simply wrong. Coming out of Italy with an excess of male hormones and ready to take the world by storm, S., failing along the way to develop a sense of balance, stumbled against some realities that he had not foreseen, and is now bitter. Since I do not know him personally and have not been involved in any of the unpleasantness that he describes, I am not prepared to join with those who call him a fraud or a con artist. But I am saddened to learn that Y., who although undoubtedly frustrated has never to my knowledge been bitter, has been drawn to S.'s institute. Although Y.'s ideas have not received much attention they have not been entirely ignored. Kip Thorne, a well known relativist at Caltech, used to say that he devoted his time to Einstein's theory during six days each week, but on Thursdays would study other theories of gravity. Y.'s was one of these. Thorne felt he had to keep an open mind because it is possible that Einstein could be wrong. However, Einstein's theory was his favorite because, aside from its elegance, it has greater predictive power (fewer adjustable constants) than most of the others. If I recall rightly, Y.'s main theory has the same predictive power as Einstein's. Most people just don't regard it as equally simple and elegant. S.'s ideas seem to me to have no grace whatever.

With best wishes,

Bryce DeWitt"

The Gravity Research Foundation Essay written by DeWitt in 1953 is another example of how he communicated with non-specialists; *see* Sect. III.3.

V.2 Research Centers

The "Centers" in the Department of Physics of the University of Texas at Austin are organized research units. The Center for Relativity was established in 1962 under the directorship of Alfred Schild who served in that capacity until 1972. At that time Bryce DeWitt became the director, and in 1987 Richard Matzner became the third (and current) director of the Center for Relativity.

A brief history of the Center for Relativity for the period 1962 to 1999 has been filed with Bryce DeWitt's archives at the Center for American History (*see* Sect. V.4). The role of the Center in numerical relativity is described in Sect. II.1).

There are, as expected, many reports requested by various agencies and the University units that address the needs and accomplishments of the Centers. I have selected one dated 5 Sept. 2004, signed by Cécile and Bryce DeWitt, because it is short, simple, and basic.

It is a reply to two questions asked by the department.

1. *Is the infrastructure necessary for research best set up on a departmental basis alone or by a combination of Research Centers and departmental support?*

   Answer:
   "The needs and resources are different for experimental and theoretical research.
   Before addressing the needs of theoretical research one may ask oneself if it would be beneficial for theoretical research to be integrated with experimental research in related areas. This pattern exists at Saclay and it might be useful to learn how theorists and experimentalists view this pattern which has existed for several years."

2. *Does the present Center structure tend to Balkanize the faculty and lead to an inequitable distribution of resources across the department?*

   Answer:
   "Three related remarks:

   1. Obviously no one would wish "equitable distribution" to mean alignment at a lower level of support.

2. Resources available. The Centers were created to attract University resources additional to basic departmental resources. These resources are designated for specific goals and cannot be pooled into a common account.
   Therefore eliminating the Centers would lead to "equitable distribution" at a lower level, and interfere with productive research.
3. "Balkanize". Was the Austro-Hungarian Empire preferable to the Balkans? One should stay clear of loaded words which are not necessarily appropriate."

Bibliography

V.1 A Biographical Memoir by Steven Weinberg

BRYCE SELIGMAN DEWITT
January 8, 1923–September 23, 2004
A Biographical Memoir[1]
By Steven Weinberg

Bryce Seligman DeWitt, professor emeritus in the physics department of the University of Texas at Austin, died on September 23, 2004. His career was marked by major contributions to classical and quantum field theories, in particular, to the theory of gravitation.

DeWitt was born Carl Bryce Seligman on January 8, 1923, in Dinuba, California, the eldest of four boys. His paternal grandfather, Emil Seligman, left Germany around 1875 at the age of 17 and emigrated to California, where he and his brother established a general store in Traver. Emil married Anna Frey, a young woman who had emigrated from Switzerland at about the same time. They had 11 children, whom Anna raised in the Methodist church.

In 1921 DeWitt's father, who had become a country doctor, married the local high school teacher of Latin and mathematics. Her ancestors were French Huguenots and Scottish Presbyterians. DeWitt was raised in the Presbyterian Church, and the only Jewish elements in his early life were the matzos that his grandfather bought around the time of Passover. DeWitt described his early exposure to religion as a boy in California in a moving memoir, "God's Rays", published posthumously in *Physics Today*. His grandmother told him that Armageddon would come in the summer of

[1] republished with kind permission by the National Academy of Sciences, Washington, D.C. (©2008).

C. DeWitt-Morette, *The Pursuit of Quantum Gravity: Memoirs of Bryce DeWitt from 1946 to 2004*, DOI 10.1007/978-3-642-14270-3_6, © Springer-Verlag Berlin Heidelberg 2011

1997 and hounded his grandfather to his deathbed, trying to make him give up his belief in Darwinian evolution.

Looking back in his memoir, DeWitt came to the conclusion that it was love that gave Christianity its overwhelming impact, but that love "needs no religious framework whatever to exert its power."

DeWitt's mother chafed at her rural surroundings and determined that her sons would live elsewhere. At the age of 12 DeWitt entered Middlesex School in Concord, Massachusetts. The headmaster at Middlesex had initiated a national scholarship program similar to the one at Harvard, and DeWitt had taken (at his mother's insistence) the competitive examination in which he earned his admission.

He graduated from Middlesex at the age of 16, and was admitted to Harvard and Caltech. He chose Harvard because he had become passionate about rowing while at Middlesex, and Harvard had "crew." He eventually stroked the Harvard Varsity. As a physics major he was deferred from military service but always felt guilty about it. Upon graduation in 1943 he went to work on the Calutron at Berkeley, the accelerator used in the Manhattan Project for the final separation of uranium-235 from uranium-238. (This had been recommended to him by Robert Oppenheimer when, because DeWitt wanted to get back to California, he had turned down Oppenheimer's invitation to join a secret research project in an undisclosed location.) He spent seven months at the Berkeley branch of the Manhattan Project and then asked to be released. He reasoned that any bright youngster could do what he was doing (hand soldering, reading meters, general gofer work), and he didn't see that his physics degree was relevant. In January 1944 he enlisted in the navy and became a naval aviator, but World War II ended before he saw combat.

DeWitt came back to Harvard in January 1946. In 1947 he began his thesis work under the nominal supervision of Julian Schwinger. The topic he chose, the quantization of the gravitational field, became his life's work. In 1949 he began his first postdoctoral year at the Institute for Advanced Study. When Wolfgang Pauli, in November of that year, learned what he was working on, he remained silent for several seconds, alternately nodding and shaking his head (a well-known Pauli trademark), and then said, "That is a very important problem. But it will take somebody really smart!"

In 1950 two major but totally unrelated developments occurred in his life. First, he became engaged to be married to Cécile Morette, a young French physicist who was in her second postdoctoral year at the institute. Second, at the urging of their father, he and his three brothers began the legal procedures for changing their surname to a name from their mother's side of the family. The younger boys were, or had been, at school in the eastern United States, and all had encountered repeated misunderstandings and false assumptions based solely on their surname, something that had seldom occurred in California.

From June to December 1950 DeWitt was with Pauli at the ETH in Zürich, and afterward he went to Bombay to spend a postdoctoral year at the Tata Institute of Fundamental Research.

This sojourn did not make good professional sense, but it suited his roving spirit. Unfortunately it ended in an abrupt and serious illness, which forced his return to Europe. In May 1951 he and Cécile were married in Paris, and in July they were in Les Houches, where the famous l'Ecole d'Eté de Physique Théorique was starting its first year. This school had been created by Cécile as a penance for marrying a foreigner, but she also saw it as something potentially valuable in its own right. It was certainly valuable to DeWitt, who during the summers he was there, was exposed to a very broad range of topics in theoretical physics. That the school was also valuable to others is attested by the fact that at its jubilee in 2001 it numbered among its past students and lecturers 26 who later became Nobel laureates and two who became Fields medalists.

In September 1951 DeWitt, this time accompanied by Cécile, returned to the Tata Institute, determined to complete his postdoctoral year. Their eldest daughter was born in Bombay in April 1952. Three other daughters were born during the following decade. In the summer of 1952 Cécile was back at Les Houches while DeWitt was looking for a job in the United States. His years abroad had kept him out of the market for academic appointments, so he accepted a job at the nuclear weapons laboratory at Livermore, where he remained for three and a half years. During his stay at Livermore, in addition to writing a treatise on "The Operator Formalism in Quantum Perturbation Theory," he became the lab's expert on (2+1)-dimensional hydrodynamical computations (impelled by NATO's desire to possess nuclear artillery shells). This expertise was applied by him years later in computations of the behavior of colliding black holes, and by his students in a variety of astrophysical problems.

Through the efforts of John Wheeler, who had become aware of his work on quantum gravity, DeWitt was offered and accepted the directorship of the Institute of Field Physics at the University of North Carolina in Chapel Hill. His initial title at UNC was visiting research professor, which enabled him to teach, or not, as he chose, and to have students. With his very first student, and with the aid of the book of Jacques Hadamard on the Cauchy problem, he discovered the basic properties of Green's functions in curved spacetime. He was also led to the beginnings of a manifestly covariant quantum theory of gravity in which, unlike the usual approach to quantum mechanics, the hamiltonian has no role to play.

In quantum mechanics the commutator AB-BA of any two quantities A and B is inferred from a quantity known as the Poisson bracket, which is calculated on the basis of classical mechanics. DeWitt came upon the 1952 paper of Rudolf Peierls, which gave a global definition of the Poisson bracket in terms of these Green's functions. Peierls's definition yields a completely unambiguous Poisson bracket for any pair of quantities, whose definition does not depend on the choice of coordinate system. The problem now addressed by DeWitt was to extend these classical results to the quantum theory with all its infinities.

In January 1957 Cécile, who had also been given the title of visiting research professor, organized the first of the general relativity and gravitation (GRG) conferences: "On the Role of Gravitation in Physics." The participants included Christian Møller, Leon Rosenfeld, Andre Lichnerowicz, Hermann Bondi, Thomas Gold, Dennis Sciama, Peter Bergman, John Wheeler, and Richard Feynman. Samuel Goudsmit had recently threatened to ban all papers on gravitation from *Physical Review* and *Physical Review Letters* because he and most American physicists felt that gravity research was a waste of time. The conference aimed to point out the shallowness of this view. In those early years, arguments were often put forward that gravity should not be quantized. Feynman vigorously disagreed and became interested in the problem while visiting Chapel Hill. Four years later, at the GRG conference in Warsaw, Feynman gave the first correct statement of how to quantize gravity (and also the non-Abelian gauge field) in the one-loop order of perturbation theory. He was the inventor of what are known as "ghosts" in non-Abelian gauge theories. These theories, invented in 1954 by Chen-Ning Yang and Robert Mills, became the subject of much of DeWitt's future work, and later turned out to furnish the basis of successful theories of all of the observed interactions of elementary particles except gravitation.

DeWitt, who had followed Feynman's work closely, extended it to two-loop order in 1964. In the meantime he had pushed forward on several other fronts. On three occasions he had presented courses of lectures at Les Houches. In 1963 he gave his most famous course, "The Dynamical Theory of Groups and Fields," which was published as a book the following year. In it he introduced a condensed notation applicable to all field theories, extended Schwinger's heat kernel methods to curved spacetime and other nonconstant backgrounds, and gave the first (and now standard) nonperturbative definition of the effective action as a Legendre transform of the logarithm of the vacuum persistence amplitude.

By the end of 1965 he had found the rules for quantizing the gravitational and non-Abelian gauge fields to all orders. But this work did not get published until late 1967 for two reasons. First, his Air Force grant was terminated and he could not pay the page charges that *Physical Review* had begun levying. Second, there seemed to be no rush. The standard model of electroweak interactions had not yet been worked out and the fundamental importance of the non-Abelian gauge field was not fully understood. Dimensional regularization, which would make renormalization easy, had not yet been invented. And he was momentarily sidetracked by John Wheeler's eagerness to develop a canonical approach to quantum gravity based on Dirac's theory of constraints. The application of Dirac's methods to gravity had some interesting features of its own. DeWitt was led to what subsequently became known as the Wheeler-DeWitt equation, which has since been applied many times to problems in quantum cosmology.

DeWitt's paper on the non-Abelian Feynman rules finally appeared two weeks before a paper by Faddeev and Popov deriving the same rules. These rules were seized upon by 't Hooft and Veltman who, apparently unaware of DeWitt's contributions, proceeded to call Feynman's ghosts "Faddeev-Popov ghosts", a name that has stuck.

In the summer of 1968 DeWitt was visited by Max Jammer, who was thinking of writing a book on the interpretation of quantum mechanics and its history. DeWitt was astonished to learn that Jammer had never heard of Hugh Everett III, who had published a paper on this topic in the same issue of *Reviews of Modern Physics* in which contributions from the 1957 Chapel Hill conference had appeared. In fact Everett's paper, which proposed that one should regard the formalism of quantum mechanics as providing a representation of reality in exactly the same sense as the formalism of classical mechanics was once thought to do, had been totally ignored by the physics community during the intervening years. DeWitt resolved to correct this situation and in 1970 wrote a popular article in *Physics Today* expounding Everett's views.

These views, although almost totally rejected at first, have little by little gained increasing numbers of adherents. The assumption that the formalism of quantum mechanics provides a direct representation of reality implies the existence of what from the point of view of classical physics would appear as many "realities." Everett's interpretation consequently became known as the "many-worlds" interpretation. DeWitt, who found Everett's ideas liberating in the sense that they lead one to ask questions that might not occur to one otherwise, became regarded as one of the foremost champions of the many-worlds interpretation, although it was always peripheral to his main interests.

By 1970 the DeWitts had begun to think of leaving Chapel Hill. Several years earlier Bryce's title had been changed to professor while Cécile had been demoted to lecturer. In addition, upon the death of Agnew Bahnson Jr., the Winston-Salem industrialist who had founded and provided financial security for the Institute of Field Physics and upon his widow's transfer of its backup funds to the university, the status of the institute underwent an abrupt change. No longer was it possible to offer postdoctoral positions with the assurance that funds would be available even if grant money failed to materialize. The postdocs of earlier years had included Felix Pirani, Ryoyu Utiyama, Peter Higgs, and Heinz Pagels. This stream of talented people had now come to an end.

In the fall of 1971 DeWitt accepted a visiting professorship at Stanford. The physics department was looking for a replacement for Leonard Schiff,

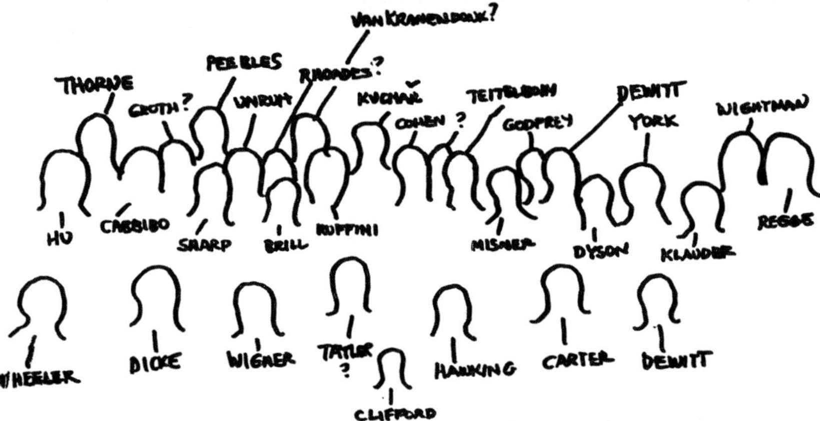

Fig. 12 The Clifford Centennial Conference (Princeton, 21 February 1970). In the fifties, gravitation and Einstein s equations were, for most physicists, on the back burner. By 1970 a few hearty souls had tackled General Relativity. Many of them can be seen in this conference picture

who had died the year before. Stanford indeed looked promising, not least because the mathematics department expressed an interest in hiring Cécile. The members of the physics department were sufficiently pleased by Bryce's visit that they made preparations to offer him a professorship. This, however, was vetoed by Felix Bloch, who upon learning that Bryce had changed his surname 20 years earlier, refused to allow the offer to proceed.

An alternative then appeared at the University of Texas at Austin. A few years earlier Alfred Schild had secured the university's agreement to establish a well-funded Center for Relativity. Schild, as its director, brought to Austin such people as Roy Kerr, Robert Geroch, and Roger Penrose. In a short time these gifted young people were snapped up by other more prestigious institutions. There was always a vacancy at the Center for Relativity, and Schild was determined to get the not-so-young DeWitts. He arranged that they would both be offered full professorships, Cécile half-time at first in the astronomy department and then later full-time in the physics department.

Mixing astronomy and relativity, the DeWitts became co-leaders of a National Science Foundation-funded eclipse expedition to Mauritania in 1973. The aim of the expedition was to repeat, with modern technology, the light-deflection observations of bygone years. This effort would not have been possible without warm cooperation between the astronomy department and the Center for Relativity.

The DeWitts were instrumental in attracting to Austin John Wheeler, who was facing compulsory retirement at Princeton. Texas gave him a center of his own to which he invited people such as David Deutsch and Philip Candelas, with whom DeWitt had become acquainted during a Guggenheim year as visiting fellow at All Souls College, Oxford, in 1975–1976.

DeWitt's early years at Texas were devoted to the colliding black hole problem and to the problems of Quantum Field Theory in curved spacetime, including the problem of the conformal or Weyl anomaly and the description of Hawking radiation. He also continued to develop his hamiltonian-free approach to Quantum Field Theory. By 1983 when he again lectured at Les Houches, he was able to set the theory of conservation laws, tree theorems, and dimensional and zetafunction regularizations completely within this framework.

In the 1980s he wrote his book *Supermanifolds*. Supermanifolds are spaces that have coordinates that anticommute (in the sense that $xy = -yx$), as well as having the ordinary sort of commuting coordinates (for which $xy = yx$). The book brought together in a systematic way a number of related but never before united topics, such as supertraces, superdeterminants, Berezin integration, super Lie groups, and path-integral derivation of index theorems. A useful topology that he introduced for integration on supermanifolds is now known by mathematicians as the DeWitt topology. A second edition of *Supermanifolds* appeared in 1991.

In 1992 DeWitt and his associates completed a lattice Quantum Field Theory study of the $O(1,2)$ nonlinear sigma model in four dimensions. This model, which bears some similarities to quantum gravity, proved to be trivial in the continuum limit.

DeWitt's last book, *The Global Approach to Quantum Field Theory* (1042 pages), was published in 2003, when he was 80 years old. It effectively sets forth his special viewpoint on theoretical physics and includes the following unique contents:

- A derivation of the Feynman functional integral from the Schwinger variational principle and a derivation of the latter from the Peierls bracket;
- Proofs of the classical and quantum tree theorems;
- A careful statement of the many-worlds interpretation of quantum mechanics in the context of both measurement theory and the localization-decoherence of macroscopic systems, which leads to the emergence of the classical world;
- A display of the many roles of the measure functional in the Feynman integral, from its relation to the Van Vleck-Morette determinant in semi-classical approximations to its justification of the Wick rotation procedure in renormalization theory;
- Repeated use of the heat kernel in a wide variety of contexts, including a zeta-function computation of the chiral anomaly in curved spacetime;
- An exhaustive analysis of linear systems, both bosonic and fermionic, and their behavior as described through Bogoliubov coefficients;
- A novel approach to ghosts in non-Abelian gauge theories: use of the Vilkovisky connection to *eliminate* the ghosts in the closed-timepath formalism that is used to calculate "in-in" expectation values; and
- A proof of the integrability of the Batalin-Vilkovisky "master" equation.

DeWitt's obituary in *Physics Today* notes that:[2] "as a scientist, Bryce was bold and extraordinarily clear thinking. He eschewed bandwagons and the common trend of trying to maximize one's publication list. Most of his papers are long masterpieces of thought and exposition. Indeed, Bryce had a rare, perfect combination of physical and mathematical intuition and raw intellectual power that was very rarely surpassed."

To this I would add that he was a fount of wisdom about theoretical physics for his colleagues at the University of Texas. His death has left a gap in our working lives that time does not seem to cure.

For his many contributions to physics DeWitt received the Dirac Medal of the Abdus Salam International Centre for Theoretical Physics (Trieste), the Pomeranchuk Prize of the Institute of Theoretical and Experimental Physics (Moscow), and the Marcel Grossmann Prize (with Cécile). Shortly before his death he was named the recipient of the American Physical Society's Einstein Prize for 2005. He was elected to membership in the National Academy of Sciences in 1990; he was also a member of the American Academy of Arts and Sciences. DeWitt was an indefatigable trekker and mountain climber, traveled widely, and lectured in many parts of the world. He is survived by his wife, Cécile, and four daughters.

---

[2] *Physics Today*, March 2005, p. 84.

This memoir incorporates materials provided to me by Bryce DeWitt before his death.

2 Publications

There are two lists of publications. The first one is Bryce DeWitt's own list with minor editing – the entries are labeled [BD, his numbering]. The second list called 'Other Writings' has been prepared by Cécile DeWitt.

Bryce DeWitt's own List of Publications

[BD 1] "Point Transformations in Quantum Mechanics", *Physical Review* **85**, 663–661 (1952) – **9th Most Cited Paper**

[BD 2] (With Cécile M. DeWitt) "The Quantum Theory of Interacting Gravitational and Spinor Fields", *Physical Review* **87**, 116–122 (1952)

[BD 3] "State Vector Normalization in Formal Scattering Theory", *Physical Review* **100**, 905–911 (1955)

[BD 4] *The Operator Formalism in Quantum Perturbation Theory*, University of California Radiation Laboratory, Publication no. 2884 (1955), 280 pp

[BD 5] "Transition from Discrete to Continuous Spectra", *Physical Review* **103**, 1565–1571 (1956)

[BD 6] "Dynamical Theory in Curved Spaces. I. A Review of the Classical and Quantum Action Principles", *Reviews of Modern Physics* **29**, 377–397 (1957) – **6th Most Cited Paper**. This article is one of the papers from the conference on the Role of Gravitation in Physics (Chapel Hill, N.C. – January, 1957)

[BD 7] "Principal Directions of Current Research Activity in the Theory of Gravitation", *Journal of Astronautics* **4**, 23–28 (1957)

[BD 8] "Principal Directions in Current Research on Gravitation", in: *Advances in Astronautical Sciences*, Vol. 1., Plenum Press, New York (1957), pp 157–168

[BD 9] "Quantum Theories of Gravity", *General Relativity and Gravitation* **1**, 181 (1958)

[BD 10] "The Scientific Uses of Large Space Ships", *General Atomic Report GAMD* 965 (1959), 40 pp

[BD 11] (With Robert W. Brehme) "Radiation Damping in a Gravitational Field", *Annals of Physics (N.Y.)* **9**, 220–259 (1960). **Selected by Bryce DeWitt as one of his most important works – 5th Most Cited Paper**

[BD 12] "Freinage dû à la Radiation d'une Particule dans un Champ de Gravitation", in: *Les Theories Relativistes de la Gravitation*, Colloque International organisé a Royaumont du 21 au 27 juin 1959 (Centre National de la Recherche Scientifique, Paris, 1962), pp 335–343. *See also* Cécile DeWitt: "Grandeurs Relatives A Plusieurs Points. Tenseurs Généralises", pp 327–333

[BD 13] "Invariant Commutators for the Quantized Gravitational Field", *Physical Review Letters* **4**, 317–320 (1960)

[BD 14] "Quantization of Fields with Infinite-Dimensional Invariance Groups", *Journal of Mathematical Physics* **2**, 151–162 (1961)

[BD 15] "Invariant Commutators for the Quantized Gravitational Field", in: Recent Developments in General Relativity. Pergamon Press, London (1962), pp 175–189

[BD 16] "Quantum Theory without Electromagnetic Potentials", *Physical Review* **125**, 2189–2191 (1962)

[BD 17] (With Ryoyu Utiyama) "Renormalization of a Classical Gravitational Field Interacting with Quantized Matter Fields", *Journal of Mathematical Physics* **3**, 608–618 (1962) – **7th Most Cited Paper**

[BD 18] "Quantization of Fields with Infinite-Dimensional Invariance Groups. II. Anticommuting Fields", *Journal of Mathematical Physics* **3**, 625–636 (1962)

[BD 19] "Definition of Commutators via the Uncertainty Principle", *Journal of Mathematical Physics* **3**, 619–624 (1962)

[BD 20] "Quantization of Fields with Infinite-Dimensional Invariance Groups. III. Generalized Schwinger-Feynman Theory", *Journal of Mathematical Physics* **3**, 1073–1093 (1962)

[BD 21] "The Quantization of Geometry", in: M. E. Rose (ed.) *Proceedings of the Eastern Theoretical Physics Conference*, October 26 & 27, 1962. Gordon and Breach, New York (1963), pp 353–386

[BD 22] "The Quantization of Geometry", in: L. Witten (ed.) *Gravitation: An Introduction to Current Research*. Wiley, New York (1963), Chap. 8, pp 266–381; this chapter has been recommended by D. Marolf for inclusion in extenso

[BD 23] (With Cécile M. DeWitt, eds.) *Relativity, Groups and Topology, 1963* Les Houches Lectures. Gordon and Breach, New York (1964), 929 pp

[BD 24] "The Quantization of Geometry", in: *Proceedings on the Theory of Gravitation*, Conference in Warszawa and Jabłonna 25–31 July 1962 (PWN-Editions Scientifiques de Pologne, Warszawa (1964), pp 131–147

[BD 25] "Gravity", in: Frederick I. Ordway, III (ed.) *Advances in Space Science and Technology*, Vol. VI. Academic Press, New York (1964) pp 1–37

[BD 26] (With Cécile M. DeWitt) "Falling Charges", *Physics* **1**, 3–20 (1964) – **Selected by Bryce DeWitt as one of his most important works**

[BD 27] "Theory of Radiative Corrections for Non-Abelian Gauge Fields", *Physical Review Letters* **12**, 742–746 (1964)

[BD 28] "Gravity: A Universal Regulator?", *Physical Review Letters* **13**, 114–118 (1964) – **8th Most Cited Paper**

[BD 29] *Dynamical Theory of Groups and Fields*. Gordon and Breach, New York (1965), 248 pp – This text first appeared in [BD 23]. Handwritten solutions are available in the archives of the UT Center for American History. They are available at http://repositories.lib.utexas.edu/. Translated into Russian and published by "Nauka", Moscow (1987)

[BD 30] "Superconductors and Gravitational Drag", *Physical Review Letters* **16**, 1092–1093 (1966) – **10th Most Cited Paper**

[BD 31] "Quantum Theory of Gravity. I. The Canonical Theory", *Physical Review* **160**, 1113–1148 (1967). Also reprinted in: L. Z. Fang and R. Ruffini (eds.) *Quantum Cosmology*, World Scientific, Singapore (1987). Translated into Russian and published by "Nauka", Moscow, 1987 – **Most Cited Paper**

[BD 32] "Quantum Theory of Gravity. II. The Manifestly Covariant Theory", *Physical Review* **162**, 1195–1239 (1967). Also reprinted in: R. N. Mohapatra and C. H. Lai (eds.) *Gauge Theories of Fundamental Interactions*, World Scientific, Singapore (1981). Translated into Russian and published by "Nauka", Moscow (1987). **Selected by Bryce DeWitt as one of his most important works – 2nd Most Cited Paper**

[BD 33] "Quantum Theory of Gravity. III. Application of the Covariant Theory", *Physical Review* **162**, 1239–1256 (1967). Also reprinted in: R. N. Mohapatra and C. H. Lai (eds.) *Gauge Theories of Fundamental Interactions.* World Scientific, Singapore (1981) – **4th Most Cited Paper**

[BD 34] "Eversion of the 2-Sphere", in: C. M. DeWitt and J. A. Wheeler (eds.) *Battelle Rencontres: 1967 Lectures in Mathematics and Physics.* W. A. Benjamin, New York (1968), pp 546–557

[BD 35] "The Everett-Wheeler Interpretation of Quantum Mechanics," in: C. M. DeWitt and J. A. Wheeler (eds.) *Battelle Rencontres: 1967 Lectures in Mathematics and Physics.* W. A. Benjamin, New York (1968), pp 318–332

[BD 36] "Spacetime as a Sheaf of Geodesics in Superspace", in: M. Carmeli, S. I. Fickler, and L. Witten (eds.) *Relativity: Proceedings of the Relativity Conference in the Midwest.* Plenum Press, New York (1970), pp 359–374

[BD 37] "Quantum Mechanics and Reality", *Physics Today* **23**(9), 30–35 (1970)

[BD 38] (With L. E. Ballentine, P. Pearle, E. H. Walker, M. Sachs, T. Koga, and J. Gerver) "Quantum Mechanics Debate", *Physics Today* **24**(4), 36–44 (1971)

[BD 39] "Quantum Theories of Gravity", *General Relativity and Gravitation* **1**, 181–189 (1970)

[BD 40] "The Many-Universes Interpretation of Quantum Mechanics", in: *Proceedings of the International School of Physics "Enrico Fermi" Course IL: Foundations of Quantum Mechanics.* Academic Press, New York (1971), pp 211–262. Also reprinted in: DeWitt and Graham (eds.) *The Many-Worlds Interpretation of Quantum Mechanics.* Princeton University Press (1973)

[BD 41] (With R. Neill Graham) "Resource Letter IQM-1 on the Interpretation of Quantum Mechanics", *American Journal of Physics* **39**, 724–738 (1971)

[BD 42] "Covariant Quantum Geometrodynamics", in: J. R. Klauder (ed.) *Magic Without Magic: John Archibald Wheeler.* W. H. Freeman, San Francisco (1972), pp 409–440

[BD 43] Book review of *The Large Scale Structure of Spacetime* by S.W. Hawking and G.F.R. Ellis, Cambridge (1973) in *Science* **182**, 705–706 (1973)

[BD 44] (With J. R. Oppenheimer) *Lectures on Electrodynamics.* Gordon and Breach, New York (1970), 164 pp

[BD 45] (With Cécile M. DeWitt, eds.) *Black Holes, 1972 Les Houches Lectures.* Gordon and Breach, New York (1973), 574 + 161 pp

[BD 46] (With Neill Graham, eds.) *The Many-Worlds Interpretation of Quantum Mechanics.* Princeton University Press (1973), 250 pp

[BD 47] (With F. Estabrook, H. Wahlquist, S. Christensen, L. Smarr, and E. Tsiang) "Maximally Slicing a Black Hole", *Physical Review* **D7**, 2814–2817 (1973)

[BD 48] (With R. A. Matzner and A. H. Mikesell) "A Relativity Eclipse Experiment Refurbished", Sky and Telescope **47**, 301–306 (1974). The original version "Report on the Relativity Experiment at The Solar Eclipse of 30 June 1973", is available at the DeWitts' archives.

[BD 49] "The Texas Mauritanian Eclipse Expedition", in: *Gravitation and Relativity: Proceedings of the 7th International Conference on General Relativity and Gravitation, Tel Aviv University*, June 1974. Keter Publ House, Jerusalem (1975), pp 81–86

[BD 50] "Quantum Field Theory in Curved Space", in: *Particles and Fields–1974: AIP Conference Proceedings No. 23, Particles and Fields Subseries No. 10.* American Institute of Physics, New York (1975), pp 660–688 – **3rd Most Cited Paper**

[BD 51] "Quantum Field Theory in Curved Spacetime", *Physics Reports* **19c**, 295–357 (1975). Translated into Russian for the series *Cherniye Diri: Novosti Fundamentalnoi Fizike.* Mir, Moscow (1978)

[BD 52] (With other members of the Texas Mauritanian Eclipse Team) "Gravitational Detection of Light: Solar Eclipse of 30 June 1973. I. Description of Procedures and Final Results", *Astronomical Journal* **81**, 452–454 (1976)

[BD 53] (With L. Smarr, A. Čadež, and K. Eppley) "Collision of Two Black Holes: Theoretical Framework", *Phys. Rev. D* **14**, 2443 (1976)

[BD 54] "Gravitational Deflection of Light. Solar Eclipse of 30 June 1973", in: G. Tauber (ed.) *Albert Einstein's Theory of General Relativity*. Crown Publ, New York (1979), pp 125–126

[BD 55] "Quantum Gravity: The New Synthesis", in: S. W. Hawking and W. Israel (eds.) *General Relativity, An Einstein Centenary Survey*. Cambridge University Press, Cambridge (1979), pp 680–745

[BD 56] (With C. F. Hart and C. J. Isham) "Topology and Quantum Field Theory", in: *Themes in Contemporary Physics: Essays in Honour of Julian Schwinger's 60th Birthday*. North-Holland, Amsterdam (1979). Also: *Physica* **96A**, 197–211 (1979)

[BD 57] "The Formal Structure of Quantum Gravity", in: M. Lévy and S. Deser (eds.) *Recent Developments in Gravitation, Cargèse 1978*. Plenum Press, New York (1979), pp 275–322

[BD 58] "Quantum Gravity", in: B. Kursunoglu, A. Perlmutter, and L. F. Scott (eds.) *On the Path of Albert Einstein*. Plenum Press, New York (1979), pp 127–143

[BD 59] "A Gauge Invariant Effective Action", in: C. J. Isham, R. Penrose and D. W. Sciama (eds.) *Quantum Gravity II*. Oxford University Press (1981), pp 449–487

[BD 60] "Approximate Effective Action for Quantum Gravity", *Phys. Rev. Letters* **47**, 1647–1650 (1981)

[BD 61] (With P. van Nieuwenhuizen) "Explicit Construction of the Exceptional Superalgebras F(4) and G(3)", *J. Math. Phys.* **29**, 1953–1963 (1982)

[BD 62] "Gravitational Deflection of Light: Solar Eclipse of June 30, 1973", in: *National Geographic Society Research Reports Vol. 14*, 1973 Projects. National Geographic Society (1982), pp 149–155

[BD 63] "The Gauge Invariant Effective Action for Quantum Gravity and Its Semi-Quantitative Approximation", in: K. Kikkawa, N. Nakanishi, and N. Nariai (eds.) *Gauge Theory and Gravitation*. Springer-Verlag, Berlin (1983), pp 189–203

[BD 64] "Quantum Gravity", *Scientific American* **249**(6), 112–129 (1983)

[BD 65] (With Raymond Stora, eds.) *Relativity Groups and Topology II*. North-Holland, Amsterdam (1984), 1323 pp

[BD 66] "The Spacetime Approach to Quantum Field Theory", in: Bryce DeWitt and Raymond Stora (eds.) *Relativity, Groups and Topology II*. North-Holland, Amsterdam (1984), pp 381–738

[BD 67] *Supermanifolds*. Cambridge University Press, Cambridge (1984), 306 pp

[BD 68] "Topics in Quantum Gravity", in: B. Bertotti, F. De Febie and A. Pascolini (eds.) *General Relativity and Gravitation*. D. Reidel, Dordrecht (1984), pp 439–451

[BD 69] *Symposium in Honor of James R. Wilson at the University of Illinois in October 1982*

(a) "The Early Days of Lagrangian Hydrodynamics at Lawrence Livermore Laboratory," in Joan M Centrella, James M. LeBlanc, Richard L Bowers (eds.) Numerical *Astrophysics: Proceedings of the Symposium*, Foreword by John Archibald Wheeler. Jones and Bartlett, Boston (1985), pp 474–481

(b) Bryce DeWitt speaking at the James Wilson Banquet (unpublished)

[BD 70] Book review of S. Chandrasekhar "*The Mathematical Theory of Black Holes*". Oxford Univ Press (1983), in: *SIAM Review* **27**, 97–99 (1985)

[BD 71] "Changing Topology", in: T. Goldman and M. M. Nieto (eds.) *Proceedings of the Santa Fe Meeting: First Annual Meeting (New Series) of the Division of Particles and Fields of the American Physical Society, October 31–November 3, 1984*. World Scientific, Singapore (1985), pp 432–436

[BD 72] "Dynamical Suppression of Topological Change", in: M. A. Markov, V. A. Berezin, and V. P. Frolov (eds.) *Proceedings of the Third Seminar on Quantum Gravity, October 23–25, 1984, Moscow, USSR*. World Scientific, Singapore (1985), pp 103–122

[BD 73] (With Arlen Anderson) "Does the Topology of Space Fluctuate?", *Foundations of Physics* **16**, 91–105 (1986)

[BD 74] "Effective Action for Expectation Values", in: R. Penrose (ed.) *Proceedings of the Quantum Gravity Discussion Conference, Oxford, March 1984*. Oxford University Press (1986), pp 325–336

[BD 75] "The Effective Action", in: I. A. Batalin, C. J. Isham, and G. A. Vilkovisky (eds.) *Quantum Field Theory and Quantum Statistics: Essays in Honour of the 60th Birthday of E. S. Fradkin*. Adam Hilger, Bristol/Boston (1987), pp 191–222

[BD 76] "The Effective Action", in: P. Ramond and R. Stora (eds.) *Architecture of Fundamental Interactions at Short Distances*. North-Holland, Amsterdam (1987), pp 1023–1057

[BD 77] "Does Conventional Quantum Gravity Exist?" in: M. Markov, V. Berezin, and V. Frolov (eds.) *Quantum Gravity: Proceedings of the Fourth Seminar on Quantum Gravity, Moscow, USSR, 25–29 May 1987*. World Scientific, Singapore (1987), pp 94–124

[BD 78] Book review of: S. W. Hawking and W. Israel (eds.) "Three Hundred Years of Gravitation. Cambridge (1987), in: *Am. J. Phys.* **55** (1988), pp 1050–52

[BD 79] "The Uses and Implications of Curved-Spacetime Propagators: A Personal View", Dirac Medal Lecture, International Centre for Theoretical Physics, Trieste (1988), pp 11–40;
Reprinted in: S. A. Fulling and F. J. Narcowich (eds.) *Forty More Years of Ramifications: Spectral Asymptotics and Its Applications (Discourses in Mathematics and Its Applications)*. Texas A & M University (1991), pp 27–48
Also reprinted in: Y. J. Ng (ed.) *Julian Schwinger. The Physicist, the Teacher, and the Man*. World Scientific, Singapore (1996), pp 33–59 (and introduced by "Preliminary Remarks" pp 29–31)

[BD 80] "The Casimir Effect in Field Theory", in: A. Sarlemijn and M. J. Sparnaay (eds.) *Physics in the Making–Essays in Honor of H. B. G. Casimir*. North-Holland, Amsterdam (1989), pp 247–272

[BD 81] "Nonlinear Sigma Models in 4 Dimensions as Toy Models for Quantum Gravity", in: S. De Filippo, M. Marinaro, G. Marmo, and G. Vilasi (eds.) *Geometrical and Algebraic Aspects of Nonlinear Field Theory*. Elsevier/North-Holland, Amsterdam (1989), pp 97–112

[BD 82] "The Vilkovisky Effective Action", in: G. Domokos and S. Kovesi-Domokos (eds.) *TeV Physics: Proceedings of the Johns Hopkins Workshop on Current Problems in Particle Theory 12*. World Scientific, Singapore (1989)

[BD 83] "Nonlinear Sigma Models in 4 Dimensions: A Lattice Definition", in: J. Audretsch and V. de Sabbata (eds.) *Quantum Mechanics in Curved Space-Time*. Plenum Press, New York/London (1990), pp 431–471; an abstract under the same title can be found in DeWitt's files.

[BD 84] (With Cécile DeWitt-Morette) "The Pin Groups in Physics", *Phys. Rev. D* **41**, 1901–1907 (1990)

[BD 85] "Nonlinear Sigma Models in 4 Dimensions as Toy Models for Quantum Gravity," in: A. Ashtekar and J. Stachel (eds.) *Conceptual Problems of Quantum Gravity: Proceedings of the 1988 Osgood Hill Conference.* Birkhäuser, Basel/Boston (1991)

[BD 86] (With E. Myers, R. Harrington, A. Kapulkin, J. de Lyra, S. K. Foong, and T. Gallivan) "Lattice Quantization of the $O(1,2)/O(2)\times Z_2$ Sigma Model in 4 Dimensions", in: M. A. Markov, V. A. Berezin, and V. P. Frolov (eds.) *Quantum Gravity (Proceedings of the Fifth Moscow Quantum Gravity Seminar).* World Scientific, Singapore (1991), pp 18–26

[BD 87] (With Eric Myers, Rob Harrington, and Arie Kapulkin) "Noncompact Nonlinear Sigma Models and Numerical Quantum Gravity", *Nuclear Physics B (Proc. Suppl.)* **20**, 744–748 (1991)

[BD 88] (With Jorge L. deLyra, See Kit Foong, T. E. Gallivan, Rob Harrington, Arie Kapulkin, Eric Myers, and Joseph Polchinski) "The Quantized $O(1,2)/O(2)\times Z_2$ Sigma Model Has No Continuum Limit in Four Dimensions. I. Theoretical Framework", *Phys. Rev.* **D46**, 2527–2537 (1992)

[BD 89] (With Jorge L. deLyra, See Kit Foong, T. E. Gallivan, Rob Harrington, Arie Kapulkin, Eric Myers, and Joseph Polchinski) "The Quantized $O(1,2)/O(2)\times Z_2$ Sigma Model Has No Continuum Limit in Four Dimensions. II. Lattice Simulation", *Phys. Rev.* **D46**, 2538–2552 (1992)

[BD 90] "How Does the Classical World Emerge from the Wave Function?" in: F. Mansouri (ed.) *Quantum Gravity and Beyond.* World Scientific (1993), pp 3–16

[BD 91] *Supermanifolds (2nd edn.)* Cambridge University Press, Cambridge (1992), 407 pp

[BD 92] "Decoherence Without Complexity and Without an Arrow of Time", in: J. Halliwell, J. Pérez-Mercader, and W. H. Zurek (eds.) *Physical Origins of Time Asymmetry.* Cambridge University Press, Cambridge (1994), pp 221–233

[BD 93] "Reminiscences of Julian Schwinger", in: Y. J. Ng (ed.) *Julian Schwinger. The Physicist, the Teacher, and the Man.* World Scientific, Singapore (1996), pp 29–56
(a) "Preliminary remarks before beginning his technical talk", pp 29–31
(b) "The uses and implications of curved-spacetime propagators: A personal view", pp 33–59. This is based on a lecture given at the International Centre for Theoretical Physics, Trieste, on the occasion of receiving the Dirac Medal. *Also in*: S. A. Fulling and F. J. Narcowich (eds.) *Forty More Years of Ramifications: Spectral Asymptotics and Its Applications – Discourses in Mathematics and Its Application, No. 1,* Department of Mathematics. Texas A& M University, College Station, Texas (1991), pp 27–48

[BD 94] (With Carmen Molina-Paris) "Gauge Theory Without Ghosts", in: C. DeWitt-Morette, P. Cartier, and A. Folacci (eds.) *Functional Integration: Basics and Applications (Cargèse, 1996).* Plenum, New York (1997), pp 327–361

[BD 95] "The Quantum and Gravity: The Wheeler-DeWitt Equation", in: *Proceedings of the Eighth Marcel Grossmann Conference, The Hebrew University, Jerusalem, Israel.* World Scientific, Singapore (1997)

[BD 96] "The Probability Interpretation of Quantum Mechanics", in: S.C. Lim (ed.) *Frontiers in Quantum Physics.* Springer, New York (1998), pp 39–49

[BD 97] "The Quantum Mechanics of Isolated Systems", *Int. J. Mod. Phys. A* **13**, 1881–1916 (1998)

[BD 98] (With Carmen Molina-París) "Quantum Gravity Without Ghosts", *Mod. Phys. Lett. A* **13**, 2475–2479 (1998)

[BD 99] "Quantum Field Theory and Spacetime – Formalism and Reality", in: T. Y. Cao (ed.) *Conceptual Foundations of Quantum Field Theory*. Cambridge University Press, Cambridge (1999), pp 176–186

[BD 100] "The Peierls Bracket", in: C. DeWitt-Morette and J.-B. Zuber (eds.) *Quantum Field Theory: Perspective and Prospective*. NATO Science Series, Kluwer Academic, Dordrecht (1999), pp 111–138

[BD 101] Book review of: David Deutsch, *The Fabric of Reality*. Penguin Press, (1997), in: *Natural Science*, May, 1998.

[BD 102] Book review of: John Archibald Wheeler with Kenneth Ford Geons, *Black Holes and Quantum Foam, A Life in Physics*. W. W. Norton (1998), in: *Physics in Perspective* **1**, 224–2251 (1999)

[BD 103] The Global Approach to Quantum Field Theory. Oxford University Press (2003, with correction 2004), 1042 pp – **Selected by Bryce DeWitt as one of his most important works, in particular Chaps: 4, 10, 24, 27, and 28; and pages 691–696, 239–243, and 258–259**. Reviews by Antoine Folacci and Bruce Jensen, in: *J. Phys. A: Math. Gen.* **36**, 12346–12347 (2003)

[BD 104] "The Everett interpretation of quantum mechanics", in: J. D. Barrow, P. C. W. Davies and C. L. Harper, Jr. (eds.) *Science and Ultimate Reality: quantum theory, cosmology, and complexity*. Cambridge University Press, Cambridge (2004) pp 167–198

[BD 105] (With Cécile DeWitt-Morette) "From the Peierls bracket to the Feynman functional integral", *Annals of Physics* **314**, 448–463 (2004)

[BD 106] Posthumous: "The Space of Gauge Fields: Its Structure and Geometry", in: G. 't Hooft (ed.) *50 Years of Yang-Mills Theory*. World Scientific, Singapore (2005), pp 15–34

[BD 107] Posthumous: "Cécile DeWitt-Morette (1922 –)", in: Nina Byers and Gary Williams (eds.) *Out of the Shadows: Contributions of Twentieth-Century Women to Physics*. Cambridge University Press, Cambridge (2006). pp 324–333. This article is an abbreviated text of the manuscript submitted on September 7, 1999 itself based on an earlier bibliography by DeWitt that appeared in: Louise S. Grinstein, Rose K. Rose, and Miriam H. Rafailovich (eds.) *Women in Chemistry and Physics: A Bibliographic Source Book*. Greenwood Publ, Westport CT (1993), pp 150–161. The full text can be found in DeWitt's archives.

[BD 108] Posthumous: "Quantum Gravity: Yesterday and Today", Edited by Cécile DeWitt-Morette and Brandon DiNunno. *General Relativity and Gravitation* **41**, 413–419 (2008). Errata **41**, 671 (2009)

Other Writings

[1] Harvard Physics Ph.D. thesis (1950):
I. The theory of Gravitational Interactions
II. The Interaction of Gravitation with Light

[2] Related to Ph.D. Thesis: "On the application of Quantum Perturbation theory to Gravitational Interactions", 1950
(a) The Einstein-Mie Theory, Spinors, The background space and the Approximation method
(b) Interaction representation, Vacuum induced stressed, Self-energies of mesons and photons

[3] Bryce Seligman "A Report on Two Papers by Professor Fermi on High Energy Nuclear Events", talk given at an International Conference on Elementary Particles at the Tata Institute of Fundamental Research in Bombay, December 14–22 (1950)
[4] "A numerical method for two-dimensional lagrangian hydrodynamics", University of California Radiation Laboratory Livermore Site UCRL 4250 Dec. 10, 1953
[5] "New Directions for Research in the Theory of Gravitation". Gravity Research Foundation Essay, New Boston N.H. 1953
[6] "Pair Production by a Curved Metric (abstract)", *Physical Review* **90**, 357 (1953)
[7] "Quantum Mechanics," unpublished lecture notes of a course given at the Ecole d'Eté de Physique Théorique (Les Houches) in 1953, available at <http://repositories.lib.utexas.edu/handle/2152/19>
[8] "Remarks on a presently neglected area of physical research", from material on the Institute of Field Physics (Chapel Hill), 1955
[9] "Dirac Theory of Constraints", publication # 2 of the Institute of Field Physics (Chapel Hill), February 1959
[10] Translation of the N. Bohr and L. Rosenfeld paper "On the question of the measurability of the electromagnetic field strengths", *Det. Kgl. Danske Videnskabernes Seskab., Mat.-fys.* **Med XII**, 8 (1933), 1960
[11] "Quantum Theory of Measurement", Lectures delivered at a theoretical physics seminar at Chapel Hill, 1968
[12] "Report on the Relativity Experiment at the Solar Eclipse of 30 June 1973", by: Bryce S. DeWitt, Richard A. Matzner, and A. H. Mikesell. Original version of [BD 48].
[13] "A Unique Effective Action for Gauge Theories", Lecture given at Les Houches, 1985
[14] Letter to Sam Schweber, December 6, 1988
[15] "Platonic Quantum Mechanics: Taking the theory literally", in: Hans H. Diebner, Timothy Druckrey, and Peter Weibel (eds.) *Sciences of the Interface*. Genista-Verlag, Tübingen (2001), pp 6–14
[16] "Comments on Martin Gardner's 'Multiverses and Blackberries' ", *Skeptical Inquirer*, March 2002, pp 60–61
[17] "God's Rays" *Physics Today*, January 2005. pp 32–34; translated into Italian "Sopra un Raggio di Luce" Di Renzo Editore, 2005. "Readers Respond to 'God's Rays' ", *Physics Today*, June 2005, pp 12–13
[18] "Probability in the deterministic theory known as Quantum Mechanics", in: G.W. Gibbons, E.P.S. Shellard, S.J. Rankin (eds.) *The Future of Theoretical Physics and Cosmology*. Cambridge University Press, Cambridge (2005), pp 667–674
[19] "Experimental Relativity and the foundations of physical theory"
[20] Extract of a referee report to the *American Journal of Physics* "Everett's Theory and the 'Many-Worlds' Interpretation"
[21] "The Everett Interpretation of Quantum Mechanics"
[22] (With C.J. Isham) "Quantum Mechanics in Strong Gravitational Fields" Draft of part of Chapter 1. University of Texas Center for Relativity preprint ca. 1983. A handwritten table of contents combining sections in B. De Witt's and C Isham's handwritings has been identified by Isham as the proposed contents of a book they were once going to write.
[23] B DeWitt and S Christensen (Ed.) *Bryce DeWitt's Lectures on Gravitation*, Lect. Notes Phys. **826**, Springer-Verlag Berlin Heidelberg (2011)
[24] "Why Physics?" (handwritten, probably mid-nineties)
[25] "Bahnson dropped a pack of camels", C. & B. DeWitt. Samples of Research at Chapel Hill **34**, 20 (1959)

Proposals and Applications (mentioned in this book)

[1] Proposal for a Lattice Quantum Field Theory Test of the ($\lambda\Phi^4$) Model in Four Dimensions.
[2] Application for the Guggenheim Fellowship

Interviews

[1] Interview by Kenneth Ford on 2/28/1995 Oral History Collection at the Center for History of Physics at the American Institute of Physics. Transcript available in full at <http://www.aip.org/history/ohilist/23199.html>
[2] Bryce DeWitt, "Sopra un raggio di luce" Di Renzo Editore 2005. Interviews with Sante Di Renzo conducted in English, translated into Italian. Preface by Ignazio Licata

Essays in honor of Bryce DeWitt

Steven M. Christensen (ed.) *Quantum Theory of Gravity, Essays in honor of the 60th birthday of Bryce S. DeWitt.* Adam Hilger, Bristol (1984). They include:

L. Smarr, "The Contributions of Bryce DeWitt to Classical General Relativity"
C. J. Isham, "The Contributions of Bryce DeWitt to Quantum Gravity"
S. A. Fulling, "What Have We Learned from Quantum Field Theory in Curved Space–Time?"
S. M. Christensen, "The World of the Schwinger–DeWitt Algorithm and the Magical a_2 Coefficient"
P. C. W. Davies, "Particles do not Exist"
P. Candelas and D. W. Sciama, "Is There a Quantum Equivalence Principle?"
L. Parker, "Some Cosmological Aspects of Quantum Gravity"
J. S. Dowker, "Vacuum Energy in a Squashed Einstein Universe"
L. H. Ford, "Aspects of Interacting Quantum Field Theory in Curved Space–Time: Renormalization and Symmetry Breaking"
J. W. York, Jr., "What Happens to the Horizons when a Black Hole Radiates?"
J. D. Bekenstein, "Black Hole Fluctuations"
R. M. Wald, "Black Holes, Singularities and Predictability"
G. Vilkovisky, "The Gospel according to DeWitt"
A. Strominger, "Is there a Quantum Theory of Gravity?"
J. A. Wheeler, "Quantum Gravity: the Question of Measurement"
W. G. Unruh, "Steps towards a Quantum Theory of Gravity"
M. R. Brown, "Quantum Gravity at Small Distance"
E. T. Tomboulis, "Renormalization and Asymptotic Freedom in Quantum Gravity"
D. G. Boulware, "Quantization of Higher Derivative Theories of Gravity"
R. Penrose, "Donaldson's Moduli Space: a 'Model' for Quantum Gravity?"
C. J. Isham, "Quantum Geometry"
J. B. Hartle and K. V. Kuchaø, "The Role of Time in Path Integral Formulations of Parametrized Theories"
C. Teitelboim, "The Hamiltonian Structure of Two-Dimensional Space–Time and its Relation with the Conformal Anomaly"
K. S. Stelle, "Supersymmetry, Finite Theories and Quantum Gravity"
P. C. West, "Properties of Extended Theories of Rigid Supersymmetry"

S. Deser, "Cosmological Topological Supergravity"
J. G. Taylor, "Future Avenues in Supergravity"
S. L. Adler, "Dynamical Applications of the Gauge-Invariant Effective Action Formalism"
R. Jackiw, "Liouville Field Theory: a Two-Dimensional Model for Gravity?"
D. Deutsch, "Towards a Quantum Theory without 'Quantization'"
L. Smolin, "On Quantum Gravity and the Many-Worlds Interpretation of Quantum Mechanics"
C. R. Doering and C. DeWitt-Morette, "The Positivity of the Jacobi Operator on Configuration Space and Phase Space"
L. Halpern, "On Complete Group Covariance without Torsion"

V.3 Curriculum Vitae

Bryce S. DeWitt (January 8, 1923 – September 23, 2004)

Birthplace: Dinuba, California

Wife's Name: Cécile Andrée Paule Morette

Children: four daughters

Education:

S.B. (Summa cum laude)	1943	Harvard University
M.A.	1947	Harvard University
Ph.D.	1950	Harvard University

Employment History:

1943	Junior Research Physicist, Radiation Laboratory, University of California, Berkeley
1944–45	U.S. Naval Aviator
1949–50	Member Institute for Advanced Study, Princeton, New Jersey
1950	Research Associate, Eidgenössische Technische Hochschule, Zürich, Switzerland
1951–52	Fulbright Research Scholar, Tata Institute of Fundamental Research, Bombay, India
1952–55	Senior Research Physicist, Radiation Laboratory, University of California, Berkeley and Livermore
1956–61	Visiting Research Professor, Department of Physics, University of North Carolina at Chapel Hill
1961–65	Professor of Physics, University of North Carolina at Chapel Hill
1965–71	Agnew Hunter Bahnson Jr. Professor of Physics, University of North Carolina at Chapel Hill

Fall 1971	Visiting Professor, Stanford University, Stanford, California
1972–87	Professor of Physics and Director, Center for Relativity, The University of Texas at Austin
1986–2004	Jane and Roland Blumberg Professor of Physics, The University of Texas at Austin

Academic[3] Awards and Fellowships:

1950	National Research Council Postdoctoral Fellow
1953	First Prize, Gravity Research Foundation
1962–	Fellow, American Physical Society
1964–66	National Science Foundation Senior Postdoctoral Fellow
1975–76, 1986	University of Texas Research Fellow
1975–76	Guggenheim Fellow
1987	Dirac Medal of the Abdus Salam International Centre for Theoretical Physics (Trieste)
1988	University of Texas Award for Outstanding Graduate Teaching
1990–	Member, National Academy of Sciences
2000	The 9th Marcel Grossman Award of the International Center for Relativistic Astrophysics, jointly with Cécile DeWitt-Morette
2003	The Pomeranchuk Prize of the Institute of Theoretical and Experimental Physics (Moscow), jointly with L.D. Faddeev
2005	The Einstein Prize of the American Physical Society

Visiting Positions and Other Short Term Appointments:

1954, 1964, 1966	Member, Institute for Advanced Study, Princeton, New Jersey
1959	Consultant, General Atomic, Division of General Dynamics Corp., La Jolla, California
1963, 1972, 1983	Directeur des Etudes, Ecole d'Eté de Physique Théorique, Les Houches, Haute Savoie, France
Fall 1964	Fulbright Lecturer, University of Osaka, Osaka, Japan
May–July 1973	Leader, Texas Mauritanian Eclipse Expedition
1975–76	Visiting Fellow, All Souls College, University of Oxford, England

[3] A nonacademic recognition, the Harvard crew, with Bryce as stroke, won the Adams Cup Regatta in 1942. DeWitt had chosen Harvard over the California Institute of Technology because of its crew.

Fall 1979	Member, Institut des Hautes Études Scientifiques, Bures-sur-Yvette, France
Spring 1980, 1981	Group Leader, Institute for Theoretical Physics, Santa Barbara, California
Spring 1982	Visiting Professor, Université de Paris, Pierre et Marie Curie
Oct.– Nov. 1988	Visiting Scientist, Joint Institute for Nuclear Research, Dubna, USSR.

Committees and Other Special Positions:

1959–71	International Committee on General Relativity and Gravitation (Secretariat, Institute of Theoretical Physics, University of Berne, Switzerland)
1970–72, 92–95	Associate Editor, Journal of Mathematical Physics
1973–	International Astronomical Union
1974–76	Executive Committee of the Division of Particles and Fields of the American Physical Society
1974–76, 80–83	National Research Council Associateships Panel, Washington, D.C.
1978–90	The Explorers Club

Special Courses of Lectures Given:

Summers 1953, 1956, 1963, 1983	Ecole d'Eté de Physique Théorique, Les Houches, Haute Savoie, France
June 1970	Université de Paris, Paris, France
July 1970	International School of Physics, "Enrico Fermi", Varenna, Italy
January 1973	Université de Marseille, France
July 1973	Oxford University, Oxford; Imperial College, London; King's College, London
May 1977	Japan Society for the Promotion of Science, Osaka University, Japan
May–July 1978	Imperial College, London
July 1978	Institut d'Etudes Scientifiques de Cargèse, Corsica
February 1982	Centro Brasileiro de Pesquisas Fisicas, Rio de Janeiro
October 1984	Academy of Sciences of the USSR
May 1986	P.N. Lebedev Physical Institute, Moscow
December 1986	University of Madras, India
May 1989	"Ettore Majorana" Centre for Scientific Culture, Erice, Sicily
September 1996	Institut d'Etudes Scientifiques, Cargèse, France

Fig. 13 1987 – Bryce DeWitt receiving the Dirac Medal of the Abdus Salam International Centre for Theoretical Physics (Trieste, Italy) for his fundamental contributions to the study of classical and quantum gravity and non-Abelian gauge theory

Fig. 14 1990 – Elected Member of the National Academy of Sciences

CONFERENCES

Pomeranchuk conference brings scientists from around the globe together in Moscow

Some 150 scientists from Russia, Ukraine, the US, Germany, Finland, Switzerland, the Czech Republic, Italy and Japan came to Moscow in January to attend the International Conference "I Ya Pomeranchuk and physics at the turn of centuries". Dedicated to the memory of Isaak Yakovlevich Pomeranchuk (1913–1966), on what would have been his 90th birthday, the conference was organized by the Institute for Theoretical and Experimental Physics (ITEP), where Pomeranchuk founded his school of theoretical particle physics, and by the Moscow Engineering Physics Institute (MEPhI), where he lectured and supervised many students. The conference covered various fields of physics that were influenced by Pomeranchuk, including high-energy physics and quantum field theory, the physics of liquid helium-3, condensed matter physics, astrophysics and cosmology, and the physics of electromagnetic processes in matter. During the conference, the Pomeranchuk Prize for 2002 was presented to Bruce DeWitt of Texas University, Austin, US (*CERN Courier* October 2002 p31).

Left to right: Nikolay Narozhny of MEPhI, Nobel prize winner Douglas Osheroff of Stanford University, Lev Okun of ITEP and Bruce DeWitt of Texas University.

The scientific director of ITEP Mikael Danilo (right), presenting the Pomeranchuk Prize for 2002 to Bruce DeWitt.

Fig. 15 2002 – Bryce DeWitt receives the Pomeranchuk Prize of the Institute of Theoretical and Experimental Physics (Moscow) jointly with L.D. Faddeev for the discovery and development of quantization methods in gauge theories, which laid the foundation for understanding the quantum dynamics of gauge fields

4 Archives Stored at the Center for American History

The Center for American History has graciously offered to be the repository for Bryce DeWitt and Cécile DeWitt-Morette's archives.

The Archives of American Mathematics (AAM) is a component of the Dolph Briscoe Center for American History at the University of Texas at Austin. No appointment is necessary to use the collections; however, they are non-circulating, so your research must be confined to the reading room. For more information about the Briscoe Center, please visit the web page: http://www.cah.utexas.edu/ Some of the AAMs collections are stored off site and require advance notice for retrieval; the finding aids (box lists) note if collections are stored off site, so please check specific finding aids for location information: http://www.cah.utexas.edu/collections/math_findingaids.php

The Bryce S. DeWitt Papers are stored on-site; no advance retrieval request is required for this collection. The collection inventory may be viewed at the following URL:
http://www.lib.utexas.edu/taro/utcah/00413/cah-00413.html

Fig. 16 They [Bryce and Cécile] became lifelong sparring partners[4]

[4] Quoted from Bryce DeWitt's obituary (D Deutsch, C Isham and G Vilkovisky Physics Today, March 2005, p. 84). This figure shows Bryce and Cécile at the occasion of the 9th Marcel Grossman meeting (July 2000, Rome) where both received jointly the Marcel Grossmann award for promoting General Relativity and Mathematics research and inventing the "summer school" concept. The photo was taken by M. Gibernau (Studio Pyrénées Ceret).

About the Author

Cécile Morette, *épouse* DeWitt, was born in Paris in 1922. She lived through World War II and took part in the reconstruction of Europe. In 1951, she created the *Ecole de Physique aux Houches* and was its Director until 1972. She is the Jane and Roland Blumberg Centennial Professor in Physics, Emerita, of the University of Texas at Austin. Throughout her scientific life, based both in France and in the US, she has written 5 monographs, published 131 research articles and review papers, and edited a further 28 books.

Her first book "*L'Energie Nucléaire*" was written in 1945, at the invitation of André George. She was then *Stagiaire au CNRS* in the Frédéric Joliot – Curie Laboratory at the Collège de France. She is the author, with Yvonne Choquet-Bruhat, of the 2-volume book "*Analysis, Manifolds and Physics*", and, with Pierre Cartier, of "*Functional Integration, Action, and Symmetries*" published in 2006.

In 1949, she taught her first major course at the *Centro Brasileiro de Pesquisas Fisicas* (Rio de Janeiro). She taught her last course in 2002 at *Sharif University of Science and Technology* (Tehran).

Information technology at first overwhelmed her. In order to survive this tsunami, and to give a hand to others in the same situation, she wrote, with the help of Chloé Marien-Casey, a simple, practical manual "*I.T. for Intelligent Grandmothers*".

In 1951, she married Bryce DeWitt. She has 4 daughters and 5 grand children.

She is disinclined to write her memoirs, but is drafting a brief autobiographical essay: "*Risks I Have Taken*".

C. DeWitt-Morette, *The Pursuit of Quantum Gravity: Memoirs of Bryce DeWitt from 1946 to 2004*, DOI 10.1007/978-3-642-14270-3, © Springer-Verlag Berlin Heidelberg 2011

Name Index